BestMasters

Springer awards „BestMasters" to the best master's theses which have been completed at renowned universities in Germany, Austria, and Switzerland.

The studies received highest marks and were recommended for publication by supervisors. They address current issues from various fields of research in natural sciences, psychology, technology, and economics.

The series addresses practitioners as well as scientists and, in particular, offers guidance for early stage researchers.

Richard Moser

Plastic Tests Plastics

A Toy Brick Tensometer for Electromechanical Characterization of Elastomers

With a Preface by Prof. Dr. Siegfried Bauer

 Springer Spektrum

Richard Moser
Linz, Austria

BestMasters
ISBN 978-3-658-10529-7   ISBN 978-3-658-10530-3 (eBook)
DOI 10.1007/978-3-658-10530-3

Library of Congress Control Number: 2015942697

Springer Spektrum

Printed on acid-free paper

Springer Spektrum is a brand of Springer Fachmedien Wiesbaden
Springer Fachmedien Wiesbaden is part of Springer Science+Business Media
(www.springer.com)

Preface

Excellent equipment is available for the measurement of stress strain curves of materials, enabling careful characterizations of elastomers, polymers, paper, metals and many other materials. But sometimes, you have the wish for a very simple accurate, easy to build, transportable and desktop size stress strain measurement system. Richard Moser shows in his best master thesis how to build such a device simply from kid's toys, together with a few common parts. He shows that the system allows for accurate measurements, comparable to that of standard laboratory equipment. Being surprisingly simple in design, it is easy to rebuild the system. Applications of the device are manifold, from research in stretchable electronics, to education and exhibitions, to attract the interest of students and visitors.

Univ.Prof. Dr. Siegfried Bauer
Department of Soft Matter Physics
Johannes Kepler University Linz, Austria
Linz, March 2015

Danksagung

In einem typischen LEGO Baukasten herrscht zu Beginn ein wirres Durcheinander. Lediglich das Bild auf der Schachtel verheißt, dass aus allen lose herumliegenden Teilen etwas Großartiges entstehen kann. So stand auch ich zu Beginn lediglich mit einer Idee im Kopf vor besagtem Durcheinander, jedoch ohne der LEGO-typischen Schritt-für-Schritt-Anleitung. Diesen Bauplan galt es zu entwerfen, und nun da er in Form dieser Arbeit abgeschlossen vor mir liegt, ist mir bewusst, dass er nicht gänzlich mein Werk ist. Er ist vielmehr Ergebnis eines Entwicklungsprozesses zu welchem viele Menschen ihren Beitrag leisteten und dabei halfen ihn Teil für Teil zusammenzusetzen. Erst durch eure Unterstützung ist aus einer simplen Idee etwas entstanden, auf das ich rückblickend stolz sein kann. Daher möchte ich an dieser Stelle all jenen meinen Dank aussprechen, die mich sowohl durch fachlichen Rat als auch menschlichen Beistand durch die Höhen und Tiefen dieses Projekts begleitet haben.

Herausgreifen möchte ich hier vor allem Prof. Dr. Siegfried Bauer und Dr. Ingrid Graz, die mir ermöglicht haben diese ungewöhnliche Idee in die Praxis umzusetzen, mich anfänglichen Skeptiker stets mit gutem Zuspruch aufs Neue zu motivieren vermochten und mich oft gekonnt aus Ideensackgassen befreiten. Weiters gilt mein Dank besonders der unbändigen Geduld meiner Laborkollegen und der gesamten SoMaP Belegschaft, die trotz oftmaliger lautstarker Beschallung meinerseits durch meisterliche und auch weniger meisterliche Werke schwermetallischer Sangeskunst stets die Contenance bewahrten und mich nicht samt meiner qualitativ niederwertigen LowFi-Yoghurtbecherlautsprecher aus dem zehnten Stock oder durch die dünnen Wände des TNF Turms warfen.

Außerdem möchte ich mich beim IPPE, allen voran Prof. Dr. Zoltan Major und DI Umut Cakmak für die Unterstützung in der Endphase der

Arbeit und der Benützung der Aramis Messeinrichtung bedanken.

Mein größtes Dankeschön gilt abschließend meiner gesamten Familie und besonders meiner Mama. Du hast die Wünsche und Anliegen deiner Kinder immer über deine eigenen gestellt und Manuel und mich bei jeder Entscheidung unterstützt. Ohne deinen ständigen Rückhalt wäre ich heute nicht da wo ich bin!

Danke!

Abstract

Electronics are currently evolving from stiff and rigid (Silicon-based) systems to flexible and even stretchable polymer solutions. Development of those "plastic- and elastic electronic circuits" strongly relies on the knowledge of the mechanical properties of suitable substrate materials and the electro-mechanical characterization of stretchable electrodes. Commercial testing devices (tensometers) for obtaining the variation of mechanical stress or electrical resistance as a function of strain are typically rather costly, heavy, and need a significant amount of lab space. Therefore, to enable reliable material testing accessible for everyone, low-cost solutions are of interest.

This work shows how to use and upgrade toy bricks, based on the LEGO® Mindstorms system, for the construction of a lightweight, low-cost and easy to reproduce tensometer. In a frugal way special "intelligent" LEGO-Mindstorms compatible sensors were designed and embedded into a LEGO Technic framework. The setup is capable of performing absolute uniaxial stretches up to 30mm, while measuring tensile forces up to 35N and electrical resistances in a range of 0-1MΩ. Several measurement ranges guarantee a high acquisition accuracy. Furthermore an intuitive LabView user-interface offers real-time tracking of the measurement along with permanent control of the device. It therefore allows for stress-strain studies along with resistance-strain measurements, hence being an ideal tool for mechanical characterization of elastomers and electromechanical studies of stretchable electrodes. In addition the rather simple design provides the opportunity to use the machine for educative purposes like project-based school classes.

The system was applied to mechanically characterize polydimethylsiloxane (PDMS) with different grades of stiffness. Additionally the application for electomechanical characterization of stretchable electrodes, based on thin metal layers on PDMS is shown. The results are compared with measurements obtained on commercial equipment, with remarkable agreement. In addition to the experimental results, the building instructions for the setup are presented. In order to further the understanding of mechanical deformations, also theoretical aspects concerning elasticity of rubber-like materials are treated in this work. In particular the general mathematical handling of material deformation and its application to elastomers for determination of the Young's modulus from stress-strain measurements is presented.

Kurzfassung

Elektronik bewegt sich weg von starren Komponenten, hin zu flexiblen und anschmiegsamen Systemen. Essentiell für die Entwicklung auf dem Sektor der "Plastik- und elastischen Elektronik" ist die genaue Kenntnis der mechanischen Eigenschaften geeigneter, dehnbarer Substratmaterialien, und der elektromechanischen Charakterisierung dehnbarer Elektroden. Handelsübliche Zugprüfmaschinen zur Aufnahme des Verhaltens mechanischer Spannung oder elektrischem Widerstand bei Deformation einer Materialprobe sind jedoch oft unhandlich, teuer in der Anschaffung und haben oft einen hohen Platzbedarf. Um diese Möglichkeit einem breiten Publikum zugänglich zu machen sind einfache, preiswerte Lösungen von Interesse.

Diese Arbeit zeigt die Konstruktion einer auf LEGO® Mindstorms basierenden Zugprüfmaschine. Die Verwendung von LEGO kombiniert exakte Ergebnisse mit einfacher Handhabung und einem unkomplizierten, kostengünstigen Nachbau. Weiters erlaubt der einfache Zugang den Einsatz im projektbasierenden Schul- und Universitätsunterricht. Die Maschine ermöglicht uniaxiale Streckungen bis 30mm und synchron dazu die Erfassung von Zugkräften bis 35N und elektrischer Widerstände im Bereich von 0-1MΩ. Verschiedene Messbereiche garantieren eine hohe Messauflösung. Eine intuitive LabView Oberfläche erlaubt die ständige Kontrolle und Konfiguration des Geräts und der durchzuführenden Messungen. Das aktuelle Setup ermöglicht daher die Durchführung von Spannungs-Dehnungsmessungen und Widerstand-Dehnungsmessungen, was die mechanische Charakterisierung von (Substrat-) Elastomeren und die elektromechanische Charakterisierung dehnbarer Elektroden zulässt.

Das Setup wurde zur mechanischen Charakterisierung von Polydimethylsiloxan mit unterschiedlichen Steifigkeiten verwendet. Weiters wurde der

Verlauf des elektrischen Widerstands mit zunehmender Dehnung von Dünn-
schichtelektroden auf PDMS gemessen und die Ergebnisse mit auf konven-
tionellen Maschinen durchgeführten Messungen verglichen und bestätigt.
Neben den experimentellen Ergebnissen wird der konkrete Aufbau des
Setups behandelt. Zum besseren Verständnis mechanischer Verformungen ist in dieser Arbeit auch die theoretische Betrachtung der Elastiz-
ität Thema. Konkret wird der grundsätzliche mathematische Formalis-
mus zur Beschreibung elastischer Verformungen und darauf aufbauend die
Ermittlung des Elastizitätsmoduls gummiähnlicher Materialien gezeigt.

Contents

List of Figures

List of Tables

1 Introduction

"I do not think there is any thrill that can go through the human heart like that felt by the inventor as he sees some creation of the brain unfolding to success... such emotions make a man forget food, sleep, friends, love, everything."

— Nikola Tesla

We are surrounded by electronics. Almost everywhere we look, we can spot an electronic device. However, most of the electronics around us follow the same concepts. For years industry and research craved for developing smaller form factors and increasing performance and speed, nonetheless always with one constant parameter: electronics are rigid. But this dogma seems to collapse, since recent research and industry demands seem to break new ground and open a new perspective for future developments. Future electronics should be flexible or even stretchable, conform to arbitrary shapes and be lightweight, but should, nevertheless, be robust and reliable. Current available developments comprehend flexible and stretchable circuitries, flexible displays as shown in Figures 1.1(a) and 1.1(b) respectively, flexible power sources, stretchable photovoltaic cells, energy-harvesting, prosthetic devices, smart skins, wearable electronics and many more e.g. [1–9].

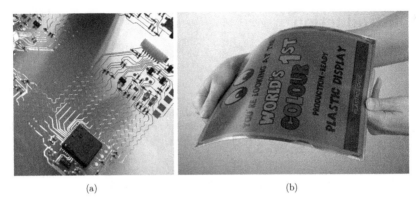

(a)   (b)

Figure 1.1 – a) Stretchable circuit fabricated using a proprietary stretchable molded interconnect (SMI) technology b) Ultra-thin, ultra-lightweight, flexible plastic display offering the prospect for a smart paper replacement[10].

Figure 1.1(a) gives an idea of the simplest case on how electronics become stretchable. By using the same components as for rigid electronics but by exchanging the rigid circuit boards and interconnects with flexible- or stretchable substrates and stretchable conductors respectively. Typically polymers and elastomers are used as substrate materials.

However, today the emphasis on energy efficiency, performance, product liability and adaptiveness are higher than ever and therefore new designs and materials demand for being thoroughly tested and characterized. Especially the (electro)mechanical properties of those products, usually displayed in work-conjugate stress-strain and resistance-strain plots, are a vital criterion for the resulting application area of a product. Material testing always needs suitable testing devices. An example of a commercially available tensometer is shown in Figure 1.2. Such machinery typically perform an one-dimensional elongation of a material sample while acquiring tensile force and strain. These stress-strain behaviour then yields material characteristic parameters like the Young's Modulus. However, in most cases such devices are rather expensive, need a significant amount of space and qualified personnel to operate the device. Nowadays, where

---

[10]Figures taken from [5] and [2].
[11]Figure taken from [10].

particularly in scientific research financial resources as well as laboratory space are often limited, new solutions for testing devices and machinery are worthwhile.

This work presents the building instructions of a "from scratch design" for a tensometer, capable of characterizing (electro)mechanical properties of common elastomers and stretchable conductors. More precisely it should allow the recording of stress-strain and resistance-strain plots with the inflicted design guidelines of keeping it simple, cost-effective, and lightweight. Nevertheless, the final device should offer a sufficient accuracy to deliver comparable and trustable results along with a certain amount of flexibility for material compatibility. Moreover it should be designed in a way, that almost everyone, that includes people who are not specially trained in mechanics or electrical engineering, are able to reproduce, understand and handle the machine. Therefore, the device could find its application on the one hand as alternative for expensive professional tensometers in scientific research, and on the other hand as demonstration device for inspiring people and students that science does not have to be complicated and expensive. It is more about having new ideas, to expect the unexpected, going for new unusual and maybe controversial approaches and most important of all: Being and staying creative.

Figure 1.2 – Professional tensile test equipment. Zwick 314 Series[11].

These catchwords perfectly describe a building platform most people are familiar with. LEGO®. These plastic bricks along with the intelligent LEGO® Mindstorms system and in combination with compatible

3rd-party and self-built sensor solutions thoroughly fulfils the mentioned requirements. Besides that as a kid I had spent many hours playing with LEGO bricks and this idea of constructing a measurement device with LEGO sounded like the perfect excuse to play with them again.

To get familiar with the necessary theory of elastic deformations the first chapter will give an overview on that matter. In particular it will define the term *elasticity* and describe the mathematical treatment of elastic mechanical deformations with the focus on rubber-like materials. Furthermore the physical background for the determination of characteristic, material specific parameters from stress-strain plots will be discussed. Being aware of the underlying theory, Section 3 will describe the specially designed LEGO-tensometer and the measurement methods in detail. The last chapter presents pertinent stress-strain measurements yielding the Young's modulus of polydimethylsiloxan (PDMS) in different constitutions along with resistance-strain measurements and performance comparison of stretchable conductors with focus on thin-film metallic conductors on elastomer substrates.

2 Fundamentals of Elasticity

"Tradition ist nicht die Anbetung der Asche sondern die Weitergabe des Feuers."

— Gustav Mahler

This chapter provides an overview on the fundamental nature and the common methods to describe the behaviour of elastic materials, in particular rubber-like materials. First the term elasticity along with different simple forms of elastic deformation is illustrated. Secondly, mathematical models and different approaches for the description of elastic characteristics of materials are shown with the focus on rubber-like materials, being capable of large elastic deformations. The last part will deal with the experimental determination of fundamental elastic properties. The presented theory is used to evaluate the experimental data in Section 4. Most of the information in this chapter is taken from the standard literature [11–17] unless otherwise stated.

The term *elasticity* fundamentally denotes to the physical property of a material to reversibly change its shape or geometrical constitution by virtue of a force acting on it. Reversibly in this case meaning that on removal of the force, the material returns to its initial condition. There are different forms of macroscopic elasticity, which are often confused upon each other, namely stretchability and flexibility or bendability. Flexible materials are substances which show the ability to reversibly change their geometrical profile by showing rather low resistance to bending or torsional forces, but relatively high resistance to tensile (stretching) forces. Stretchable materials do not have this limitation, wherefore those kind of materials in addition can also be deformed due to rather low tensile forces; they react with rather low resistance to tensile strain. Figure 2.1 gives an impression of different forms of mechanical deformation.

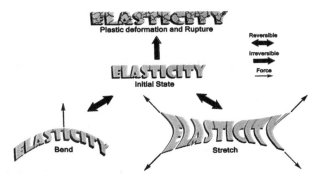

Figure 2.1 – Reversible and irreversible forms of deformation due to applied forces in elastic materials.[12]

Generally speaking, every material is elastic, or shows elastic behaviour to some extend. What discriminates commonly called elastic from inelastic materials is their degree of *reversible* deformation, in particular the amount of strain before plastic (irreversible) deformation and/or rupture occurs. Furthermore the force which is necessary to gain a significant deformation differs greatly throughout the materials. The specific deformation behaviour of a material under an applied load is distinguished by various elastic moduli, such as the Young's (or elastic) modulus and shear modulus, both applying for tensile and shear deformation respectively. Figure 2.2(a) shows the Young's moduli for different material classes.

Experimentally the elastic properties of a material are determined by tensometers capable of measuring the strain (the degree of deformation) and the opposing force while deforming a sample in a predefined way. The result of such measurements is usually presented in the form of *stress-strain* curves, which yield information about the opposing force of a material upon deformation. More precisely the average internal force per unit area, the mechanical stress σ along an axis with respect to the strain, the deformation in relation to the initial length of the material $\epsilon = \Delta l / l_0$ along

---

[12]Sketch inspired by [18].

this axis, is studied. This relation then yields information about characteristic mechanical material parameters. Figure 2.2(b) shows typical stress-strain relations for different kinds of materials under simple extension. This measurement method is also used in the LEGO-tensometer dealt with in the next chapter.

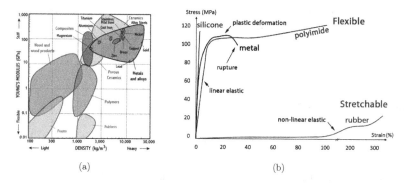

(a)   (b)

Figure 2.2 – a) Collection of Young's moduli E for different material classes b) Qualitative stress-strain curves for silicone and metal showing a very narrow (linear) area of reversible strain followed by plastic deformation and rupture. Polyimide acts as an example for a flexible material with a high Young's modulus and a rather extensive region of plastic deformation. Rubber-like materials show a rather low Young's modulus along with a broad (non-linear) elastic range.[13]

The elastic- or Young's modulus E is an important parameter of a material informing about the stiffness of an elastic material. More precisely it terms the slope of the initial stress-strain behaviour. This regime is in most cases linear and therefore complies with Hooke's Law for one dimensional stretch as stated in Equation 2.1. This relation is often sufficient for the treatment of elastic behaviour of most solids.

$$\sigma = E \cdot \epsilon \qquad (2.1)$$

As depicted by Figure 2.2(b) rubber-like materials or elastomers show a wide *non-linear* elastic range. Therefore, the latter equation does not apply for the whole elastic range. The treatment of such materials demands

---

[13]Sketches taken from [19] and [18].

for more sophisticated models.

In the following sections the methodology for determining strain and stress for arbitrary materials and deformations is laid out. The "material" in this case is assumed to be of isotropic and homogeneous nature. The microscopic structure is not yet relevant at this stage, but will become important in Section 2.3. Afterwards several approaches of modelling the elastic behaviour of rubber and the theoretical basics for determining the Young' Modulus of rubber from stress-strain measurements is presented.

2.1 The Deformation Tensor

For material tests often more complicated deformations than simple extension e.g. simple shear are used. In such cases the material deformation cannot be argued by pure geometrical considerations. To further the understanding of material deformation, the underlying theory of mechanical stress and strain with the focus on rubber-like materials is given.

Under means of external forces an arbitrary material of volume V_0 is deformed to a certain extent, therefore changing its shape and volume. In doing so, an arbitrary but fixed point inside the material in the undeformed state, indicated by the radius-vector $\vec{r}$ is shifted to a new position $\vec{r'}$ as shown in Figure 2.3. The coordinate system should remain fixed.

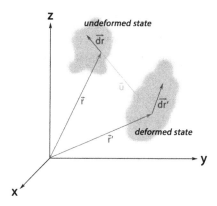

Figure 2.3 – Transformation of a point $\vec{r} \to \vec{r}'$ and infinitesimal line elements $d\vec{r} \to d\vec{r}'$ upon deformation in an arbitrary homogeneous and isotropic material. The coordinate system $\vec{e}_i \ i \ \epsilon \ \{x, y, z\}$ remains fixed during deformation.

This position shift $\vec{r} \to \vec{r}'$ can be expressed by a motion vector transformation function $\chi(\vec{r}, t)$, but it is common to denote it simply by $\vec{r}'(\vec{r}, t)$. Since stationary conditions are assumed, the time-dependence can be neglected and the point transformation can be written as

$$\vec{r}' = \chi(\vec{r}) = \vec{r}'(\vec{r}) \tag{2.2}$$

Since the deformation of the material should be determined, the transformation of an arbitrary line-element $dr \to dr'$ rather than the pure shift of points in the material is of interest.

$$\begin{aligned}
d\vec{r}' &= \chi\left(\vec{r} + d\vec{r}\right) - \chi(\vec{r}) \stackrel{\text{TaylorExp.}}{=} \chi(\vec{r}) + \frac{\partial \chi(\vec{r})}{\partial \vec{r}} d\vec{r} - \chi(\vec{r}) \\
&= \frac{\partial \chi(\vec{r})}{\partial \vec{r}} d\vec{r} \stackrel{(2.2)}{=} \frac{\partial \vec{r}'(\vec{r})}{\partial \vec{r}} d\vec{r} := \mathbf{F} \ d\vec{r} \tag{2.3}
\end{aligned}$$

$$\to \quad F_{ij} = \frac{\partial r_i'}{\partial r_j} \tag{2.4}$$

$$r_i' = F_{ij} r_j \tag{2.5}$$

$$dr_i' = F_{ij} dr_j \tag{2.6}$$

The second order tensor $\mathbf{F}$, the *Deformation Gradient Tensor* transforms an infinitesimal line element $d\vec{r}$ as well as a point $\vec{r}$ into their deformed states $d\vec{r}'$ and $\vec{r}'$ respectively. It is important to denote that $d\vec{r}'$ is associated with an error in the order of $(d\vec{r})^2$ due to neglecting higher order terms in the Taylor expansion in Equation 2.3. The tensor $\mathbf{F}$ therefore only maps deformations in the immediate vicinity of $\vec{r}$.

To be precise $\mathbf{F}$ is a general transformation in space and can always be split up into a rotational $\mathbf{R}$ and a deformation tensor $\mathbf{D}$. The orthogonal tensor $\mathbf{R}$ is a measure of the rotation of $\vec{r}$, $\mathbf{D}$ accounts for local material deformation around $\vec{r}$. To make the transformation unique $\mathbf{D}$ has to be symmetric.

$$\mathbf{F} = \mathbf{RD} \tag{2.7}$$

The length of the the initial line element $dl = |d\vec{r}|$ and the line element in the transformed state $dl' = |d\vec{r}'|$ can be calculated from Equation 2.3.

$$
\begin{aligned}
dl'^2 = d\vec{r}'^2 \overset{(2.3)}{=} & \ (\mathbf{F}\,d\vec{r})\,(\mathbf{F}\,d\vec{r}) \\
= & \ d\vec{r}\,\mathbf{F}^T\mathbf{F}\,d\vec{r} := d\vec{r}\,\mathbf{C}\,d\vec{r}
\end{aligned}
\tag{2.8}
$$

$$
\rightarrow \quad C_{jk} = F_{ij}\,F_{ik} = \left(\frac{\partial r_i'}{\partial r_j}\frac{\partial r_i'}{\partial r_k}\right) \tag{2.9}
$$

$$
dl_i'^2 = dr_j\,C_{jk}\,dr_k \tag{2.10}
$$

As shown in Equation (2.8) and (2.10) the second order tensor $\mathbf{C}$, the *Right Cauchy Green Strain Tensor* (RCGT) yields the squared length of the line element dl'^2 in relation to the square of its length dl^2 in the initial state. This relation parameter, the strain or more precisely the relative length λ^2 can be calculated as follows with $\hat{d\vec{r}}$ representing a unit vector pointing into the direction of $d\vec{r}$.

$$
\lambda^2 := \frac{dl'^2}{dl^2} \overset{(2.8)}{=} \hat{d\vec{r}}\,\mathbf{C}\,\hat{d\vec{r}} \tag{2.11}
$$

Since the rotation tensor $\mathbf{R}$ is orthogonal ($\mathbf{R}^T\mathbf{R} = 1$) it can be seen, that $\mathbf{C}$ only gives information about the deformation of the material and not about any rotation during motion. The relative length λ^2 therefore is a measure for the degree of material deformation.

$$\mathbf{C} \overset{(2.8)}{=} \mathbf{F}^T\mathbf{F} \overset{(2.7)}{=} \left(\mathbf{D}^T\mathbf{R}^T\right)(\mathbf{RD}) = \mathbf{D}^T\mathbf{D} \qquad (2.12)$$

The length of a line element remains undeformed, stretched or compressed for $\lambda = 1$, $\lambda > 1$ and $0 < \lambda < 1$ respectively.

The tensor $\mathbf{C}$ is positive definite and symmetric, which implies that it has positive Eigenvalues $\widetilde{C}_j$. This is a very important fact, since if $\mathbf{C}$ is diagonalized, corresponding to a transformation into the principal coordinate system, the diagonal elements of $\mathbf{C}$, its Eigenvalues correlate with the squares of the principal strains $\lambda_j{}^2$ in the direction of the orthogonal principal axes, which are given by the Eigenvectors $\hat{\vec{E}}_j$ of $\mathbf{C}$. Normalisation of the Eigenvectors finally yields the principal coordinate system in which only the pure strains λ_j occur. Thus $\mathbf{C}$, with respect to the new base vectors in terms of principal coordinates can be written as

$$C_{jk_{diag}} = \delta_{jk}\widetilde{C}_j \qquad (2.13)$$
$$\lambda_j{}^2 = \widetilde{C}_j \qquad (2.14)$$

The tensor $\mathbf{C}$ is a second rank tensor, and has as any of these three invariants I_i, which remain unchanged under rotation. They are given based on the diagonalised form of $\mathbf{C}$ and will become important in the next chapter where certain models for rubber-elasticity are discussed.

$$
\begin{aligned}
I_1 &= Tr(\mathbf{C}) = \lambda_1^2 + \lambda_2^2 + \lambda_3^2 \\
I_2 &= \frac{1}{2}\left(Tr(\mathbf{C}^2) - \left(Tr(\mathbf{C}^2)\right)^2\right) = \frac{1}{\lambda_1^2} + \frac{1}{\lambda_2^2} + \frac{1}{\lambda_3^2} - 3 \\
I_3 &= Det(\mathbf{C}) = \lambda_1^2\lambda_2^2\lambda_3^2
\end{aligned} \qquad (2.15)
$$

Often it is worthwhile to have strain parameters yielding the value 0 in the undeformed state. In such cases the relative stretch ε rather than the relative length λ is of interest. Since the λ_j are already known, the principal strains ε_j can be calculated rather easily.

$$\varepsilon_j \quad := \quad \frac{dl'_j - dl_j}{dl_j} = \frac{dl'_j}{dl_j} - 1 \tag{2.16}$$

$$\overset{(2.11)}{=} \lambda_j - 1 \overset{(2.14)}{=} \sqrt{\widetilde{C}_j} - 1 \tag{2.17}$$

The length of a line element remains undeformed, stretched or compressed for $\varepsilon = 0$, $\varepsilon > 0$ and $\varepsilon < 0$ respectively.

A second approach, which is often very useful to determine the relative strains and relative lengths, is to introduce the displacement vector $\vec{u}$, as indicated in Figure 2.3, rather than to perform a pure point transformation as done above.

$$\vec{r'} = \vec{u} + \vec{r} \tag{2.18}$$

Then the RCGT transforms into the following form.

$$C_{jk} \overset{(2.10)}{=} \frac{\partial r'_i}{\partial r_j} \frac{\partial r'_i}{\partial r_k}$$

$$\overset{(2.18)}{=} \frac{\partial(u_i + r_i)}{\partial r_j} \frac{\partial(u_i + r_i)}{\partial r_k} = \left(\frac{\partial u_i}{\partial r_j} + \delta_{ij}\right)\left(\frac{\partial u_i}{\partial r_k} + \delta_{ik}\right)$$

$$= \frac{\partial u_k}{\partial r_j} + \frac{\partial u_j}{\partial r_k} + \frac{\partial u_i}{\partial r_j}\frac{\partial u_i}{\partial r_k} + \delta_{jk} \tag{2.19}$$

$$:= 2U_{jk} + \delta_{jk} \tag{2.20}$$

$$\rightarrow 2U = \frac{\partial u_k}{\partial r_j} + \frac{\partial u_j}{\partial r_k} + \frac{\partial u_i}{\partial r_j}\frac{\partial u_i}{\partial r_k} \tag{2.21}$$

$$= \frac{1}{2}\left(\mathbf{C} - \mathbb{I}\right) \tag{2.22}$$

This transformation reveals $\mathbf{U}$, the *Green-Lagrange Strain Tensor*, which is similarly to $\mathbf{C}$ symmetric and positive definite, therefore, as well implying positive Eigenvalues $\widetilde{U}_j$. Assuming the tensor has already been transformed into the principal coordinate system the length of the line element in the deformed state dl' can be determined as follows.

$$dl'^2 \stackrel{(2.10),(2.20)}{=} dr_j \left(2U_{jk} + \delta_{jk}\right) dr_k$$

$$= \left(2\widetilde{U}_j + 1\right) dl_j{}^2 \tag{2.23}$$

The relative stretches ε_j and relative lengths λ_j in the direction of the principal axes can be calculated.

$$\epsilon_j \stackrel{(2.17)}{=} \sqrt{2\widetilde{U}_j + 1} - 1 \tag{2.24}$$

$$\lambda_j \stackrel{(2.11)}{=} \sqrt{2\widetilde{U}_j + 1} \tag{2.25}$$

In the case of small deformations $\varepsilon_j \to 0$, valid for most solids, also the displacement vector $\vec{u}$ tends to be small. Therefore, in Equation 2.21 differentials of second order can be neglected and the latter equations can be simplified, since $\sqrt{2\widetilde{U}_j + 1} \approx 1 + \widetilde{U}_j$. In this particular case the Eigenvalues of $\mathbf{U}$ correspond with the principal relative strains $\varepsilon_j = \widetilde{U}_j$ and the tensor $\mathbf{U}$ itself may be simplified.

$$U_{jk} \stackrel{(2.20)}{=} \frac{1}{2} \left(\frac{\partial u_k}{\partial r_j} + \frac{\partial u_j}{\partial r_k} \right) \tag{2.26}$$

As shown in this chapter for most deformation problems it is worthwhile to determine, according to the particular problem either the Green-Lagrange or the Cauchy-Green strain tensor. In the case of common (crystalline) solid materials, being only capable of small deformations mostly the Green-Lagrange tensor is used. In rubber-physics, where rather large deformations compared to classical solids are common, the use of the Cauchy-Green tensor is favourable, since its Eigenvalues directly yield the principal deformations λ_j, which are also valid for large deformations. In the next section the stress should be examined.

2.2 The Stress Tensor

If a body of a certain homogeneous and isotropic material is not deformed, meaning no external forces are acting on it, it remains in thermodynamic and mechanical equilibrium, implying that also the microscopic constitution is in an equilibrium state. Thus, an infinitesimal volume element dV can be considered to be representative for the whole body. In thermal equilibrium the resultant force acting on dV is zero. If the body is deformed, this equilibrium state is distorted and internal mechanical stresses within the material evolve, trying to return the body to its original shape.

The deforming force $\vec{f}$ can be expressed as a sum of all forces acting on every infinitesimal volume fraction dV of the body and can therefore be written as an integral over the force-density $\widetilde{\mathbf{F}}$[14].

$$f_i = \int \widetilde{\mathbf{F}}_i \, dV \qquad (2.27)$$

These infinitesimal forces can only act at the surface of the considered volume element. Thus, the latter equation can also be expressed as an area-integral, which implies that the components of the force-density $\widetilde{\mathbf{F}}_i$ can be written as the divergence of a second order tensor σ_{ij}. This allows, by use of the Gaussian divergence theorem, to write Equation 2.27 as an area-integral, with the surface elements dA_j.

$$\widetilde{\mathbf{F}}_i \quad := \quad \frac{\partial \sigma_{ij}}{\partial r_j} = \nabla \cdot \sigma_i \qquad (2.28)$$

$$f_i \overset{(2.27)}{=} \int \frac{\partial \sigma_{ij}}{\partial r_j} \, dV = \oiint_{\partial V} \sigma_{ij} \, dA_j \qquad (2.29)$$

---

[14]Parameters in relation to a volume are expressed with a $\sim$

The tensor σ_{ij}, giving the acting force per area, is the *Stress Tensor*. It is apparent form the latter equation, that $\sigma_{ij}\, dA_j$ is composed of the orthogonal stresses $\delta_{ij}\sigma_{ij}$ and the tangential stresses σ_{ij} with $i \neq j$, thus, yielding the force components f_i. Orthogonal stresses or *tensile stresses* are often denoted by σ_{ii}, tangential stresses or *shear stresses* by τ_{ij} as indicated in Figure 2.4.

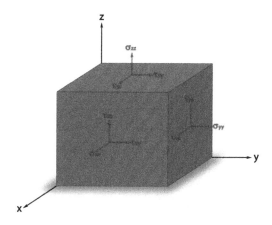

Figure 2.4 – Unit cube with volume dV of elastic material with marked tensile σ_{ii} and shear τ_{ij} stresses. The first index always identifies the surface on which the force acts, the second index gives the direction of the force.

To relate the change in shape of a volume element dV under deformation, resulting in the mentioned stresses σ_{ij}, the work W necessary to perform the deformation needs to be calculated. Therefore, the displacement vector $\vec{u}$, introduced in Equation 2.18 is used. Since it gives the displacement of a point $\vec{r} \to \vec{r}'$ the deformation work results in

$$W = \int \delta\widetilde{W}\, dV = \int \mathbf{F}_i\, \delta u_i\, dV \overset{(2.28)}{=} \int \frac{\partial \sigma_{ij}}{\partial r_j}\, \delta u_i\, dV \qquad (2.30)$$

This yields, by use of partial integration and the use of the Gaussian divergence theorem

$$\int \delta\widetilde{W}\, dV = \oiint_{\partial V} \sigma_{ij}\, \delta u_i\, dA_j - \int \sigma_{ij} \frac{\partial u_i}{\partial r_j}\, dV \qquad (2.31)$$

If an infinite body is considered, which remains undeformed at its borders, σ_{ij} vanishes and, therefore, the first term in the latter equation can be ignored. In addition as stated in Section 2.1 the Lagrange-Green strain tensor is symmetric, whereby Equation 2.31 can be written as follows

$$\int \delta\widetilde{W}\,dV = -\frac{1}{2}\int \sigma_{ij}\,\delta\left(\frac{\partial u_i}{\partial r_j} + \frac{\partial u_j}{\partial r_i}\right)\,dV \overset{(2.26)}{=} \int \sigma_{ij}\,\delta U_{ij}\,dV \quad (2.32)$$

.

$$\rightarrow \delta\widetilde{W} = -\sigma_{ij}\,\delta U_{ij} \overset{(2.22)}{=} -\frac{1}{2}\sigma_{ij}\delta C_{ij} \quad (2.33)$$

The change in work density $\delta\widetilde{W}$ necessary to perform a deformation according to δU_{ij} is, therefore, given by the stress tensor σ. To determine the stress tensor, the first and second law of thermodynamics with the inner energy U, the heat Q and the work W is needed. Furthermore the Helmholtz free energy F is of importance.

$$\delta Q = T\,dS \quad (2.34)$$
$$dU = \delta Q + \delta W = T\,dS + \delta W \quad (2.35)$$
$$dF = dU - d(TS) = \delta W - S\,dT \quad (2.36)$$

The increase of the free energy density $d\widetilde{F}$ upon deformation, if assuming that the deformation is performed under isothermal conditions $dT = 0$, matches the increase of the work density $\delta\widetilde{W}$. From this the stress tensor σ_{ij} giving the stresses of the volume element dV can be determined[15].

$$d\widetilde{F}\big|_T = \delta\widetilde{W}\big|_T \quad (2.37)$$

$$\rightarrow \sigma_{ij} \overset{(2.33)}{=} -\left(\frac{\partial \widetilde{F}}{\partial U_{ij}}\right)\bigg|_T = -2\left(\frac{\partial \widetilde{F}}{\partial C_{ij}}\right)\bigg|_T \quad (2.38)$$

---

[15]To be precise also internal stresses due to hydrostatic pressure p, leading to an additional pdV in the external work δW would have to be considered. Since most experiments are carried out at standard conditions ($p \approx 10^5 Pa$) and only the deformation due to direct applied forces should be studied, this term can be neglected in most cases. Moreover, as stated later in this chapter, rubber-like materials are often considered as incompressible, therefore $dV = 0$

Hence, due to diagonalization of $\mathbf{U}$ and $\mathbf{C}$ the latter equation can be rewritten in terms of the principal strains ε_i and λ_i, yielding the principal stresses σ_i for the volume element dV in the direction of the principal axes $\vec{e}_i$.

$$\sigma_i \stackrel{(2.14)}{=} -\left(\frac{\partial \widetilde{F}}{\partial \varepsilon_i}\right)\Bigg|_T \stackrel{(2.25)}{=} -\left(\frac{\partial \widetilde{F}}{\partial \lambda_i}\right)\Bigg|_T \qquad (2.39)$$

If the free energy density $\widetilde{F}$ is known, the internal stresses can be calculated[16]. Furthermore, it can be seen from Equation 2.36 that the deformation energy δW is distributed into an *energetic* and an *entropic* part.

$$\sigma_i = \underbrace{\left(\frac{\partial \widetilde{U}}{\partial \lambda_i}\right)\Bigg|_T}_{energetic} - \underbrace{T\left(\frac{\partial \widetilde{S}}{\partial \lambda_i}\right)\Bigg|_T}_{entropic} \qquad (2.40)$$

For most materials one of this two terms is dominating, therefore, a material can be classified as either *energy-elastic* or *entropy-elastic*. In order to categorize a material basically consideration of the underlying thermodynamics is sufficient. However, to further the understanding of the underlying mechanisms causing either energy- and entropy-elasticity, a look at the microscopic/atomic constitution of different materials is beneficial.

2.3 Microscopic Constitution

In general the elastic behaviour of a material arises due to two distinct mechanisms at the microscopic level.

The first accounts for intermolecular forces, in particular atomic bonding forces. Application of a tensile force leads to a disturbance of the equilibrium resulting in a change of the atomic distance in a bond. A simplified picture of this matter is given in Figure 2.5. Stretching the atomic bonds leads to an significant increase in potential energy and, therefore, in in-

---

[16]From this point on the simplified notation with the principal strains λ_i is used.

ternal energy U^{17}. If the overall atomic arrangement itself remains almost unchanged, therefore, the change in entropy upon deformation $dS \approx 0$, the material counts as *energy-elastic*. Materials dominated by this form of elasticity are classical solids with a regular atomic structure i.e. crystals, glasses in the condensed state or ceramics.

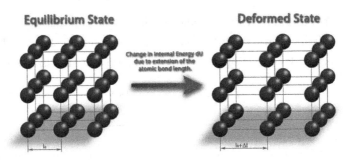

Figure 2.5 – Uniaxial extension of a simplified crystalline solid. A cuboid with unit cell length l_0 is extended to a unit cell length $l_0 + \Delta l$ resulting in a change of potential energy dU in the atomic bonds.

It is important to note that this simple scheme is only valid for small deviations from the equilibrium state[18], where the potential energy of an atomic bond with respect to its length can be assumed to be parabolic. According to Equation 2.38 this approximation leads to a linear stress-strain relation known as *Hooke's Law* in its general form.

$$\sigma_{ij} = c_{ijkl} \cdot U_{kl} \qquad (2.41)$$

For larger extensions higher order terms would become more prominent and the latter equation would be no longer be valid. However atomic bonding forces are typically of rather low range, therefore only small deformations are possible before plastic deformation and rupture arise.

---

[17]In the simplest case modelled with a Lennard-Jones potential consisting of an attractive and repulsive term $U(l) = -\frac{A}{l^6} + \frac{B}{l^{12}}$ with the atomic distance l [20].

[18]Approximation is valid for small strains $\varepsilon_{ij} \to 0$.

The second mechanism resulting in macroscopic elastic behaviour is, contrary to the first, a change in arrangement of the molecules resulting in a significant change in entropy S. In such materials the interatomic distances remain almost unchanged upon deformation, therefore, the change in internal energy $dU \approx 0$. Materials showing this kind of behaviour are counted as *entropy-elastic* and the most prominent representatives are the ideal gas and rubber-like materials like elastomers.

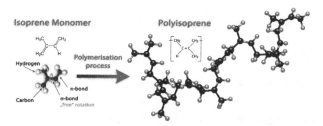

Figure 2.6 – Formation of a polyisoprene chain due to polymerization of isoprene. The n in the chemical structure of the polymer denotes the number of monomers in the chain and therefore the degree of polymerisation. The "random" chain orientation arises due to almost free rotation around σ-bonds[19].

As shown in Figure 2.6 rubber-like materials consist of chain-like molecules so-called polymers, which itself consist of, as the name depicts a covalently bonded sequence of the same monomer. Polymer chains have a covalently bonded backbone and monomer specific functional side groups, according to the monomer used. The mostly hydrocarbon backbone is hold together by σ (single)- or π (double) C-C bonds. Where the alignment of the π bonds remains rather fixed, the σ bonds allow almost free rotation among them[20]. As a consequence the structure of a polymer chain is highly variable rather than fixed and it allows for an almost random constitution of the polymer chain. This further allows for many different configurations of the backbone, while maintaining the bond length and angles constant. Thus the internal energy U remains constant in good approximation dur-

---

[19]Original sketches taken from [21–24].

[20]More precisely, backbone chain bonds show local potential energy minima confining the possible bonding angles. Nevertheless, assuming free rotation is a sufficient approximation for simple modelling of rubber elasticity as shown in Section 2.4

ing deformation.

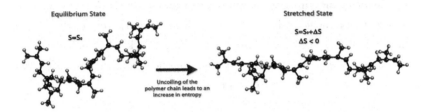

Figure 2.7 – Stretching an elastomer leads to an uncoiling of the (randomly oriented) polymerchains and therefore a decrease in entropy ΔS.

The elastic behaviour of an elastomer arises due to the uncoiling of those long-chain molecules as depicted in Figure 2.7. If the end-to-end length of the polymer is far from its fully extended state there are many different configurations accessible, all of them compatible with the fixed bond lengths and bond angles. Increasing of the end-to-end length (=stretching) reduces the number of possible configurations until in the fully extended state only one configuration remains. In other words: the degree of order in the system increases as the chain in oriented. Considering the uncertainty interpretation of entropy, as used in Section 2.4, this disentanglement accounts for a change in entropy S and may be described only with statistical methods. Since the intermolecular forces between two polymer chains, mostly Van-der-Waals [21], are rather weak compared to the covalent bonds in the backbone, also in a polymer assemblage the assumption of free rotation of the σ bonds remains a good approximation.

---

[21]The intermolecular interaction is highly depended on the sort of the side groups of the monomer.

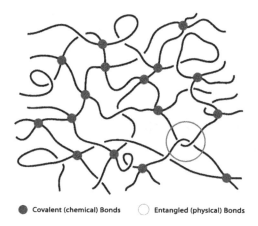

● Covalent (chemical) Bonds   ○ Entangled (physical) Bonds

Figure 2.8 – 3-dimensional polymer network. The individual chains are connected at certain points either via covalent chemical cross-links or by inherent chain entanglement[22].

In order to form a solid material, the polymer-chains must be connected at certain points, also called cross-links. This can be achieved either by a sufficient inherent entanglement, crystallisation effects or glassy domains (physical bonding), or covalent bonds (chemical bonding) to other polymers. Both forms lead to the formation of a 3-dimensional coherent network as shown in Figure 2.8. The most common method for a chemical bonding process is the vulcanization process of natural rubber, whereas the number of cross-links per unit volume defines the stiffness of the rubber; the more cross-links the stiffer the material. The length of the polymers between two cross-links still remains sufficiently large to allow the approximation of almost free (rotational) movement of the individual chains. Hence, by applying a tensile force the network deforms similarly by unfolding the polymer chains as shown in Figure 2.7.

In order to describe the elastic behaviour of such materials, in particular the stresses σ_i depicted in Equation 2.40 and therefore the change in entropy upon deformation has to be determined. A simple but yet effective statistical model for this matter is the *Neo-Hooke-Model*, which should

---

[22]Original sketch taken from [25].

be briefly outlined in the following section. Based on this and also by phenomenological approaches more refined models developed over time. In particular the *Mooney-Rivlin* as well as the approaches by *Gent* and *Ogden* are briefly exposed.

2.4 Neo-Hooke, A Statistical Approach to Rubber Elasticity

This model is thoroughly discussed and developed in [14, p.42ff] and [13, p.198ff]. Nevertheless, the basic principles and thoughts to determine the entropy change of a polymer network upon deformation will be given in the following chapter. To determine the stress-strain relation, the entropy density $\tilde{S}$ of a polymer network is required. First the entropy S_C of a single polymer chain is calculated and subsequently the results are used to widen the model to a 3D-polymer assemblage.

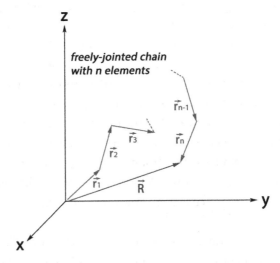

Figure 2.9 – Formation of a freely-jointed-chain starting at the origin O by randomly adding n Kuhn-segments $\vec{r}_i$ of length $a = |\vec{r}_i|$ with $1 \leq i \leq n$. This results in a random-chain with a fully-stretched length $L = na$ and an end-to-end vector $\vec{R}$ with length $R = |\vec{R}|$.

A polymer chain, as already depicted in the latter section is built up by a large number of monomers forming a covalently bonded carbon backbone with the possibility of rotation around σ-bonds. In order to determine the entropy of such a configuration the probability for a certain arrangement of the monomers or more precisely the probability distribution $p(\vec{R})$ for a distinct end-to-end vector $\vec{R}$ is needed. A rather simple but yet effective model for the build up of the polymer backbone is to approximate it as a *freely-jointed-chain*. It is assumed that the backbone consists of n links, each of length a. Those Kuhn-links typically model segments of "straight-bonded" monomers[23], whereas the number of monomers per segment depends on the type of monomer. The direction of each link is randomly distributed in space and indicated by the vector $\vec{r_i}$. The simplest treatment for modelling such a chain is to use a 3D-random-walk model with each step being independent of the previous steps. The probability distribution $p_1(\vec{r_1})$ of the first step, starting at the point of origin is

$$p_1(\vec{r}) = N\delta(r-a) \quad \text{with} \quad \int p_1(\vec{r})\,d\vec{r} \stackrel{!}{=} 1 \rightarrow N = \frac{1}{4\pi a^2} \qquad (2.42)$$

with the normalisation factor N. Adding another segment leads to a end-to-end vector $\vec{R} = \vec{r_1} + \vec{r_2}$.

$$p(\vec{R}) = \int p_1(\vec{r_1})\,p_2(\vec{R} - \vec{r_1})\,d\vec{r_1} \qquad (2.43)$$

The probability $p(\vec{R})\,d\vec{R}$ to find the end in the interval $\vec{R} + d\vec{R}$ is given by the sum of the probabilities of all possible two-step walks arriving there.

---

[23]Those planar *trans*-rotational states are favoured configurations of the skeletal backbone-bonds in n-alkanes.

In order to add a finite number of steps it is convenient to work with the
the Fourier Transform $P(\vec{\rho})$ of 2.43.

$$P(\vec{\rho}) \;=\; \int e^{-i\vec{\rho}\vec{R}} p(\vec{R}) \, d\vec{R} \tag{2.44}$$

$$\overset{(2.43)}{=} \int e^{-i\vec{\rho}\vec{R}} p_1(\vec{r_1}) \, p_2(\vec{R} - \vec{r_1}) \, d\vec{r_1} d\vec{R} \tag{2.45}$$

$$= \int e^{-i\vec{\rho}(\vec{r_1}+\vec{r_2})} p_1(\vec{r_1}) \, p_2(\vec{r_2}) \, d\vec{r_1} \, d\vec{r_2} = P_1(\vec{\rho}) \cdot P_2(\vec{\rho}) \tag{2.46}$$

This may be easily extended to n chain segments, yielding the full Fourier Transform of the complete chain. Since equal chain segments are assumed, the relation can be further simplified.

$$P(\vec{\rho}) = \prod_{i=1}^{n} P_i(\vec{\rho}) = (P_1(\vec{\rho}))^n \tag{2.47}$$

with the inverse

$$p(\vec{R}) = \frac{1}{(2\pi)^3} \int e^{-i\vec{\rho}\vec{R}} P(\vec{\rho}) \, d\vec{\rho} \tag{2.48}$$

This yields the probability distribution $p(\vec{R})$ for finding the end-vector at $\vec{R}$, which now may be used on the initial probability distribution for a single step given in Equation 2.42. The Fourier Transform is evaluated by using spherical coordinates, by taking ρ in the direction of the polar axis and with $r = |\vec{r}|$. Subsequently the Fourier Transform of the full chain with n segments $P(\vec{\rho})$ and its back-transformation $p(\vec{R})$ can be calculated.

$$P_1(\vec{\rho}) = \frac{1}{4\pi a^2} \int e^{-i\rho r \cos(\theta)} \delta(r - a) \, r^2 sin(\theta) \, dr d\phi d\theta = \frac{sin(\rho a)}{\rho a} \tag{2.49}$$

$$P(\vec{\rho}) \;=\; \left(\frac{sin(\rho a)}{\rho a} \right)^n \tag{2.50}$$

$$p(\vec{R}) \;=\; \frac{1}{(2\pi)^3} \int e^{i\vec{\rho}\vec{R}} \left(\frac{sin(\rho a)}{\rho a} \right)^n d\vec{\rho} \tag{2.51}$$

With $\vec{R}$ serving as the polar axis Equation 2.51 can be solved analogously to Equation 2.49. Moreover by substituting $\rho = \frac{y}{a\sqrt{n}}$, Equation 2.50 can be further simplified by Taylor expansion under the assumption of $n \gg 1$.

$$\left(\frac{sin(\frac{y}{\sqrt{n}})}{\frac{y}{\sqrt{n}}}\right)^n \approx \left(1 - \frac{y^2}{6n} + O[y]^4\right)^n \approx \left(e^{-\frac{y^2}{6n}}\right)^n = e^{-\frac{y^2}{6}} \qquad (2.52)$$

This yields

$$p(\vec{R}) \overset{(2.52)}{=} \frac{1}{2\pi^2 R} \int\limits_0^\infty \sin\left(R\frac{y}{a\sqrt{n}}\right) e^{-\frac{y^2}{6}} \, d\rho \qquad (2.53)$$

which can be determined analytically resulting in

$$p(\vec{R}) = \frac{b^3}{\pi^{\frac{3}{2}}} e^{-b^2 R^2} \quad \text{with} \quad b^2 = \frac{3}{2na^2} \qquad (2.54)$$

The probability distribution remains spherically symmetric, since it is only dependent on the absolute value R of the end-to-end vector $\vec{R}$. It shows an analogy to a Gaussian error function, yielding that it is most probable that both ends are at the same point, in the considered case the point of origin.

However, the most probable end-to-end distance is not zero. It is true, that the origin is the most probable *point* in space, but since the probability distribution is spherically symmetric all points in a spherical shell yield the same end-to-end distance and therefore have to be considered equally possible. The probability to find the end of the chain in a shell with radius $R + dR$ is

$$p^*(R)\, dR = \frac{4b^3}{\pi^{\frac{1}{2}}} R^2\, e^{-b^2 R^2}\, dR \qquad (2.55)$$

which yields that the probability to find the end at the origin equals zero and shows a minimum at the most probable end-to-end length

$$R^* = \frac{1}{b}. \qquad (2.56)$$

The *mean-square* value $\overline{R^2}$ will be important later in this discussion

$$\overline{R^2} = \frac{\int\limits_0^\infty R^2 p^*(R)\, dR}{\int\limits_0^\infty p^*(R)\, dR} = \frac{3}{2b^2} \qquad (2.57)$$

Now that the probability distribution of the freely and randomly jointed chain is known, the calculation of the chain entropy can be done. According to Boltzmann's Equation 2.58 the entropy depends on the number of possible configurations A, which also accounts for the probability for a certain configuration with end-to-end vector $\vec{R}$.

$$
\begin{aligned}
S &= k_B ln(A) & (2.58) \\
S_C(\vec{R}) &= k_B\, ln\left(p(\vec{R})\, d\vec{R}\right) \\
&\overset{(2.55)}{=} k_B\, ln\left(\frac{b^3}{\pi^{\frac{1}{2}}}\, d\vec{R}\right) - k_B b^2 R^2 \\
&\overset{!}{=} c - k_B b^2 R^2 & (2.59)
\end{aligned}
$$

The latter equation yields the entropy of a randomly-jointed polymer chain with n segments of length a. The factor k_B is the Boltzmann constant, c denotes a constant. Since mainly entropy changes ΔS_C are of interest, this constant is of little interest. Now it is also proven mathematically, that an expansion or contraction of the polymer-chain corresponding to a change in end-to-end distance ΔR results in a change ΔS_C and thus in a restoring force.

To determine the elastic properties of a macroscopic elastomer body, now these chains have to be joined together into a 3D network as depicted in Figure 2.8. These chains, as treated before, are defined to be the segment between two cross-links with $\widetilde{N}$ chains per unit volume and have the length depicted by Equation 2.57 in the equilibrium state. Their junction points should deflect in the same way as the macroscopic body is deformed, which allows to use the principal strains λ_i with $i \in \{1, 2, 3\}$ introduced in Section

2.1. Use of this notation on a single chain in the network deflecting from $\vec{R}_0 \to \vec{R}$ on deformation

$$S_C(\vec{R}) = c - k_B b^2 \sum_{i=1}^{3} \lambda_i^2 R_{0_i}^2 \qquad (2.60)$$

allows to calculate the change of the chain-entropy ΔS_C.

$$\Delta S_C(\vec{R}_0 \to \vec{R}) = -k_B b^2 \sum_{i=1}^{3} (\lambda_i^2 - 1) R_{0_i}^2 \qquad (2.61)$$

It should also be assumed, that the overall entropy S is the sum of the individual chain-entropies S_C in the body. With N_C being the number of polymer-chains in the considered body. It should be noted that a uniform distribution of the polymer-chains is postulated, meaning that no direction in space is favoured.

$$\sum_{i=1}^{3} \sum_{j=1}^{N_C} R_{0_{ij}}^2 = \frac{1}{3} R_{0_j}^2 = \frac{1}{3} N_C \overline{R_0^2} \qquad (2.62)$$

$$\Delta S = -k_B b^2 \sum_{i=1}^{3} \sum_{j=1}^{N_C} (\lambda_{ij}^2 - 1) R_{0_{ij}}^2 \overset{(2.62)}{=} -\frac{1}{3} k_B b^2 N_C \overline{R_0^2} \sum_{i=1}^{3} (\lambda_i^2 - 1) \qquad (2.63)$$

$$\overset{(2.57)}{\to} \Delta S = -\frac{1}{2} N_C k_B \sum_{i=1}^{3} (\lambda_i^2 - 1) \qquad (2.64)$$

For a more general formulation of the latter equation the number of chains N_C is replaced with the chain-density $\widetilde{N}$. This yields the entropy-density change $\Delta \widetilde{S}$ from which the deformation-work-density can be calculated. The change in internal energy is neglected as described before, therefore $dU = 0$. Furthermore, isothermal conditions are still valid.

$$\Delta \widetilde{F}(\lambda_i, T) \overset{(2.37)}{=} \Delta \widetilde{W}(\lambda_i, T) \overset{(2.35)}{=} -T \Delta \widetilde{S} \overset{(2.63)}{=} \frac{1}{2} \widetilde{N} k_B T \sum_{i=1}^{3} (\lambda_i^2 - 1) \qquad (2.65)$$

The constant coefficient $\widetilde{N}k_BT$ corresponds to the shear-modulus G, as shown in Section 2.6 where the use of this and the models presented in the next sections on simple forms of deformation is shown.

$$\Delta\widetilde{W} = \frac{1}{2}G\sum_{i=1}^{3}(\lambda_i^2 - 1) \tag{2.66}$$

This is known as the *Neo-Hooke-Model*. It yields the the strain-energy function of a 3D polymer network, describing the non-linear elastic properties of elastomers for small strains. The application of the model to certain simple problems, such as uniaxial stretch is important for this work, since the designed stretching device performs a uniaxial extension of the specimen to determine the elastic modulus. This rather simple model depending only on one material parameter, the chain-density $\widetilde{N}$ yields a solid basement for the understanding of the nature of rubber-elasticity. For large uniaxial strains the predictions become rather inaccurate, which will be shown in the experimental section 4.1. For simple-shear deformations the model is accurate also for larger deformations as shown in [14, p.93].

In the next chapter some more refined models for the description of rubber-elasticity are given, but not in the extent as the latter. It should rather give an impression of the ideas behind the models along with the individual results, which are again the *strain-energy-functions* $\Delta\widetilde{W}(\lambda_i, T)$.

2.5 Refined Models of Rubber Elasticity

2.5.1 The Mooney-Rivlin Approach

Contrary to the Neo-Hooke approach the Mooney-Rivlin formulation is a more phenomenological approach. It focusses on identifying a general form for the mathematical description of rubber-elasticity rather than explaining the reasons from which those phenomena arise. The basis for this formulation were given by Mooney in 1940, some years before the development of the Neo-Hooke model [26]. With some simple assumptions like the incompressibility of rubber and Hooke's Law for simple-shear extensions

Mooney derived the strain-energy function given in Equation 2.67 which turned out to be fairly similar to the later Neo-Hookean approach. More precisely the Neo-Hooke model is a special case of the Mooney model for $C_2 = 0$.

$$\Delta \widetilde{W}(\lambda_i, T) = \sum_{i=1}^{3} C_1 \left(\lambda_i^2 - 1\right) + C_2 \left(\frac{1}{\lambda_i^2} - 1\right) \quad \text{with} \quad 2C_1 = \widetilde{N}k_B T \quad (2.67)$$

The model proofed to be fairly accurate up to stretches around 10-20% and is, therefore, adequate for most applications and will amongst other models also be used in the experimental section for the determination of the Young's Modulus. The shear-modulus can be derived from the constants

$$G = 2\left(C_1 + C_2\right) \tag{2.68}$$

However, the model yielded some severe inconsistencies with experiments. Especially wrong predictions for uniaxial extension along with compression made this interpretation rather inadmissible for being a general theory describing the behaviour of rubber under various forms of stretching.

To overcome those inconsistencies Rivlin presented a theory based on Mooney's calculations [27]. Rivlin, as well as Mooney faced the problem by pure mathematics without any reference to the molecular structure. Rivlin assumed the homogeneousness and isotropy of the material in the unstrained state, requiring the strain-energy function $\widetilde{W}$ to be symmetrical with respect to the principal extensions λ_i. Furthermore, the change of sign of λ_i should not alter the strain-energy-function[24]. This implies that only even powers of λ_i occur. The three simplest forms of such rotational invariants are those of the diagonalised Right Cauchy Green Tensor (RCGT) shown in Equation 2.15. Since the material is assumed to be

---

[24]This would correspond to a rotation through 180°

incompressible $I_3 := 1$ and the strain-energy-function can be expressed in terms of I_1 and I_2.

$$\Delta \widetilde{W}(I_1, I_2) = \sum_{i=0,j=0}^{\infty} C_{ij} \left(I_1 - 3\right)^i \left(I_2 - 3\right)^j \qquad (2.69)$$

Since the terms in brackets are usually small, the first components of the sum predominate. If only the fist terms with C_{01} and C_{10}[25] are considered, Equation 2.69 corresponds to Equation 2.67, only considering C_{01} yields the result of the Gaussian-theory shown in Equation 2.66.

The latter equation does not directly provide material parameters such as the Young's or the shear-modulus, it is merely a powerful method for curve-fitting with numerous parameters C_{ij} and, therefore, allowing the determination of the stress-strain relation according to Equation 2.38 to be carried out to the desired degree of accuracy. Based on the Rivlin-model several formulations arose, mostly for specific applications and materials [14, p.231]. One of them, a version of Rivlin and Sauders, was advanced by Gent and Thomas known as the *Gent-Model*.

2.5.2 The Gent Model

The Mooney-Rivlin model, in the simple form given in Equation 2.67, yields the stress-strain relations in good agreement for small strains, in the case of uniaxial extension also for moderately large strains. Elastomers become harder for large stretches commonly referred to as *strain-stiffening*. This effect can be explained by crystallization effects and by a high degree of uncoiling of the polymer-chains which start to approach their extension limits. A scheme of this effects is shown in Figures 2.10 and a detailed discussion on this topic is found in [14, p.16ff].

---

[25]$C_{00} = 0$, since $\Delta \widetilde{W}$ has to vanish in this particular case.

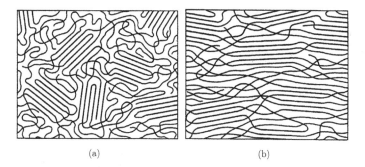

(a)   (b)

Figure 2.10 – Crystallisation effects in unstrained a) and strained b) rubber showing highly parallel oriented chains, which conform even more in the strained state, leading to a significant stiffening of the material as the maximum degree of uncoiling is reached.[26].

The models described earlier lack of this feature and, therefore, fail when strain stiffening becomes significant. To account for those effects, on speculation of Thomas the *Gent-Model* [28] introduces a logarithmic term kicking in at a certain strain and leading to a rapid ascent of $\Delta\widetilde{W}$. The parameter J_m indicates the maximum strain before rupture of the material occurs, which is given by the discontinuity of the logarithm. The Gent-Model, similar to the ones stated before, is also only valid for isotropic and homogeneous materials.

$$\Delta\widetilde{W}(I_1, I_2, J_m) = -C_1 J_m \, ln\left(1 - \frac{I_1 - 3}{J_m}\right) + C_2 \, ln\left(\frac{I_2}{3}\right) \qquad (2.70)$$

The Gent-Model needs three fitting parameters C_1, C_2 and J_m, whereby two of them are related to the shear-modulus as depicted by Equation 2.71. The coefficient C_1 is similarly to the models presented earlier related to the degree of polymerisation $\widetilde{N}$, whereas C_2 reflects physical restraints on the molecular level. The third parameter J_m is based on the stretch invariant I_1 yielding the rupture conditions

$$G = 2\left(C_1 + \frac{C_2}{3}\right) \qquad (2.71)$$

$$J_m = I_{1_m} - 3 = \lambda_{1_m}^2 + \lambda_{2_m}^2 + \lambda_{3_m}^2 - 3 \qquad (2.72)$$

---

[26]Sketch taken from [14, p.19].

where λ_{i_m} predict the maximum stretch ratios before rupture of the material occurs. In most publications the Gent-Model is found in a fairly reduced form, which is also found more often in experimental data fitting i.e. [28–30]. This approximation is based on the fact, that at moderate strains the value of I_2 is often large enough so that the second term in Equation 2.70 can be neglected, reducing it to

$$\Delta\widetilde{W}(I_1, I_2, J_m) = -C_1 J_m \, ln\left(1 - \frac{I_1 - 3}{J_m}\right) \quad \text{with} \quad G = 2C_1 \qquad (2.73)$$

The reduced form of the Gent-Model needs only two fit-parameters G and J_m and can, therefore, be pronounced as the simplest model describing the elastic behaviour of elastomers in good approximation over the whole strain range.

2.5.3 The Ogden Model

The refinement of Rivlin, makes use of the Cauchy-invariants to determine the strain-energy function and with a fair number of fitting-parameters the model yields results which are in good approximation with experiments. However, especially for higher strains this model gets rather complicated.

Ogden claimed the use of the invariants may be an unnecessary complication and, therefore, abandoned the use of those. In addition the use of only the principal strains λ_i should further the mathematical simplicity. The *Ogden-Model* also allows uneven, yet non-integer and negative powers of λ_i, reasoned by the fact that only even powers have no physical relevance [31].

$$\Delta\widetilde{W}(\lambda_i) = \sum_{i=1}^{3}\sum_{n}^{N}\frac{\mu_n}{\alpha_n}\left(\lambda_i^{\alpha_n} - 1\right) \quad \text{with} \quad G = \frac{1}{2}\sum_{n}^{N}\mu_n\alpha_n \qquad (2.74)$$

The model also accounts for strain-stiffening effects and yields the shear-modulus G as a characteristic material constant and may be extended to the degree of accuracy needed for a particular application. Reasonable results are achieved with $n \geq 3$ according to [32]. For the data-fitting in

the experimental section a 3rd order fit was performed with six fit parameters. For lower values of n and specific values of α_n the Neo-Hooke and
Mooney-Rivlin can be found as special cases of the Ogden-Model.

In addition to the models presented in this work many more exist, each
of them suitable for particular problems. One of the most sophisticated
models, the Ogden-Model still is a pure mathematical description rather
than an explaining theory, taking no reference of the microscopic constitution of elastomers into account. It works fine, but a general theory
fundamentally based on physical properties and parameters does not exist
so far.

Now that the necessary theory of the elastic properties stress and strain
along with suitable models for the elastic behaviour of rubber are estab-
lished, the application on simple problems describing the measurements
taken with the later presented stretching device is shown in the next paragraph.

2.6 Application of Rubber Elasticity to Common Forms of Deformation

For material testing the most common forms of deformation to determine certain material parameters of elastomers such as the shear- and the
Young's modulus are uniaxial and biaxial deformation and application of
shear stress (simple- and pure-shear). Therefore, the material is placed in
a device capable of deforming the material and measuring the rate of deformation and the force needed to perform the deformation. In the present
work uniaxial extension is of further interest, and is discussed in detail.

Figure 2.11 shows the simplest form of deformation, uniaxial extension.
In this form of deformation the tensile force acts perpendicular on the side
of the material, which is assumed to be of cuboid form in the equilibrium
state.

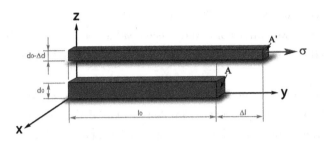

Figure 2.11 – Uniaxial extension as the simplest form of elastic deformation. A cuboid of length l_0 is stretched to a length $l_0 + \Delta l$ due to the force/stress acting on the y-surface with initial cross-section A. The sample is mechanically fixed in the x-z plane.

The deformations are pure stretches along the coordinate axes. The deformation gradient tensor and the RCGT can be written as

$$\mathbf{F} = \begin{pmatrix} 1 + \frac{\Delta d}{d_0} & 0 & 0 \\ 0 & 1 + \frac{\Delta l}{l_0} & 0 \\ 0 & 0 & 1 + \frac{\Delta d}{d_0} \end{pmatrix} \qquad (2.75)$$

$$\mathbf{C} \overset{(2.8)}{=} \mathbf{F}^T \mathbf{F} = \begin{pmatrix} (1 + \frac{\Delta d}{d_0})^2 & 0 & 0 \\ 0 & (1 + \frac{\Delta l}{l_0})^2 & 0 \\ 0 & 0 & (1 + \frac{\Delta d}{d_0})^2 \end{pmatrix} \qquad (2.76)$$

yielding the Eigenvalues λ_i^2 and the principal strains λ_i along the diagonal of $\mathbf{C}$ in terms of the orthonormal Eigenvectors E_i.

$$E_1 = (1,0,0)^T \quad E_2 = (0,1,0)^T \quad E_3 = (0,0,1)^T \qquad (2.77)$$

The determination of the principal strains is trivial in this case, since the material is deformed along the coordinate axes by a pure tensile force. Thus, solving of the Eigensystem of $\mathbf{C}$ is not particularly necessary. It can also be determined by pure geometrical considerations from Figure 2.11 that the principal strains λ_x, λ_y and λ_z correspond to the deformations along the coordinate axes, which are in accordance with the Eigenvectors of $\mathbf{C}$. For more complicated deformation, such as simple shear a pure geometrical argumentation is not trivial and, therefore, the determination of the Eigensystem is necessary.

An important constraint for elastomers is the assumption of incompress-ibility of the material class. Incompressibility means, that on deformation of the material, there is no change in volume. This is given by the determinant of the deformation gradient tensor **F**. It also leads to an important relation in terms of the principal stresses λ_i.

$$\frac{V}{V_0} = det\,(\mathbf{F}) = \prod_{i=1}^{3} \lambda_i := 1 \qquad (2.78)$$

If a sample is extended in the direction of the applied force, the material reacts with a contraction in the directions perpendicular to the applied load. The ratio of this effect is given by the *Poisson-Ratio* ν

$$\frac{\Delta d}{d_0} = \nu \frac{\Delta l}{l_0} \qquad (2.79)$$

The Poisson-Ratio is closely related to the bulk modulus K and the Young's modulus E, shown in the following equation

$$\nu = \frac{1}{2} - \frac{E}{6K} \qquad \text{with} \qquad K = -V \frac{dp}{dV} \qquad (2.80)$$

with the pressure p and the volume V. For an ideal incompressible material $dV = 0$ which leads to $K \to \infty$ and therefore yields $\nu = 0.5$. Typical elastomers like natural rubber show bulk modulus in the GPa range while having a Young's modulus in the MPa range. Therefore, the second term in the latter equation can be neglected in most cases, which justifies the assumption of incompressibility for this material class.

For large strains the Poisson-contraction causes a significant change of the cross-section $A \to A'$ on which the applied force acts. For small strains, prevailing in solid state theory, this effect is often neglected, therefore, $A \approx A'$. In hyperelastic theories, this effect cannot be neglected, wherefore two quantities are defined namely the stress per strained area A' or the *true-stress* σ_T and the stress per unstrained area A or the *nominal-*

stress.

Considering Equation 2.78 the problem can be reduced to a single λ in the extension direction, since the strains in the other directions remain symmetric.

$$1 + \frac{\Delta l}{l_0} = \lambda_y =: \lambda \quad 1 + \frac{\Delta d}{d_0} = \lambda_x = \lambda_z \stackrel{(2.78)}{=} \frac{1}{\sqrt{\lambda}} \qquad (2.81)$$

With this simplification the strain energy functions presented in the latter Section reduce to

$$I_1 = \lambda^2 + \frac{2}{\lambda} \qquad (2.82)$$

$$I_2 = 2\lambda + \frac{1}{\lambda^2} - 3 \qquad (2.83)$$

$$I_3 = 1 \qquad (2.84)$$

$$\mathbf{Neo - Hooke}: \Delta\widetilde{W}(T,\lambda) = \frac{1}{2}G(I_1 - 3) \qquad (2.85)$$

$$\mathbf{Mooney - Rivlin}: \Delta\widetilde{W}(T,\lambda) = C_1(I_1 - 3) + C_2 I_2 \qquad (2.86)$$

$$\mathbf{Gent}: \Delta\widetilde{W}(T,\lambda) = -\frac{1}{2}G J_m \ln\left(1 - \frac{I_1 - 3}{J_m}\right) \quad (2.87)$$

$$\mathbf{Ogden}: \Delta\widetilde{W}(T,\lambda) = \sum_n^N \frac{\mu_n}{\alpha_n}\left(\lambda^{\alpha_n} + \frac{2}{\lambda^{\frac{\alpha_n}{2}}} - 3\right) \quad (2.88)$$

Hence, the constitutive equation 2.39 yielding the nominal stress σ_N in the y-direction can be formulated.

$$\sigma_N \stackrel{(2.39)}{=} -\left(\frac{\partial \Delta\widetilde{W}(\lambda)}{\partial \lambda}\right)\Bigg|_T \qquad (2.89)$$

Since principal stresses are pure stresses along their principal axes, every stress is caused by a principal tensile force f_i with $\sigma_{N_i} = \frac{f_i}{A_i}$ while A_i being the initial cross-section and A_i' the cross-section in the stretched state where f_i acts on. The true-stresses σ_{T_i} transform into the nominal stresses σ_{N_i} as follows

$$\sigma_{T_i} = \frac{f_i}{A_i'} = \frac{f_i}{\lambda_j \lambda_k A_i} \overset{(2.78)}{=} \lambda_i \frac{f_i}{A_i} = \lambda_i \sigma_{N_i} \tag{2.90}$$

Although the true-stresses yield the actual stress in the material, the nominal stresses are preferably used in stress-strain measurements since the tensile force is measured during the stretch. The corresponding nominal stretch can thus be calculated rather easy since the initial cross-section is of course a constant.

$$\textbf{Neo} - \textbf{Hooke}: \quad \sigma_N = G\left(\lambda - \frac{1}{\lambda^2}\right) \tag{2.91}$$

$$\textbf{Mooney} - \textbf{Rivlin}: \quad \sigma_N = 2C_1\left(\lambda - \frac{1}{\lambda^2}\right) + 2C_2\left(-\frac{1}{\lambda^3} + 1\right) \tag{2.92}$$

$$\textbf{Gent}: \quad \sigma_N = G\,J_m\frac{\frac{1}{\lambda^2} - \lambda}{(I_1 - (J_m + 3))} \tag{2.93}$$

$$\textbf{Ogden}: \quad \sigma_N = \mu_n\left(\lambda^{\alpha_n - 1} - \lambda^{-\frac{\alpha_n}{2} - 1}\right) \tag{2.94}$$

Those principal stresses will be used in the experimental section for fitting the obtained stress-strain relations. The fit-curves in the latter equations subsequently yield the shear-modulus G which directly leads to the Young's modulus E.

$$E = 2G(1 + \nu) \tag{2.95}$$

This mathematical formalism applies for all forms of deformation. In principle the deformation gradient tensor has to be found, from which the RCGT can be calculated. This tensor yields the principal strains and their individual strain directions by solving its Eigensystem. With this the presented models can be applied. They yield the shear modulus and therefore also the parameter of interest, the Young's modulus.

3 Design of the Measurement Device

"The Dark Ages are the time between when you put away the Legos for the last time as a kid, and [when] you decide as an adult that it is okay to play with a kid's toy."

— Hillel Cooperman

Now that the necessary theory is clear, this chapter gives an overview of the measurement method and the actual build-up of the setup used. The approaches for the sample deformation along with the measurement and acquisition methods of the needed quantities for recording stress-strain and resistance-strain data are presented in detail.

The device performs an uniaxial stretch of an elastomer sample while synchronously measuring the strain and either resistance or tensile force. Therefore, the device allows for the determination of the Young's modulus for mechanical elastomer characterisation and electromechanical performance studies of stretchable conductors respectively. The goal for the design of the stretcher itself however was to build it low-cost and also to allow the duplication with only basic knowledge in mechanical or electrical engineering. However, although a simple approach had to be taken to meet those requirements, the setup should still be capable of delivering accurate and trustworthy results. LEGO-Technic® in combination with LEGO-Mindstorms® seemed to be the perfect compromise in terms of simplicity, cost-effectiveness, mechanical stability, adaptiveness and versatility.

Figure 3.2 gives an overview on the basic design of the device and Figure 3.1 gives an impression of the device itself.

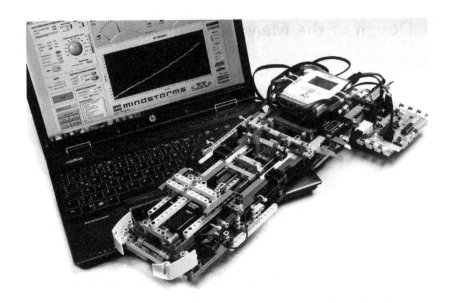

Figure 3.1 – The complete measurement setup

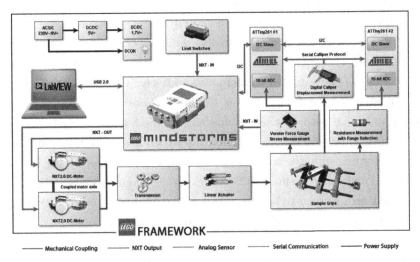

Figure 3.2 – Block diagram of the device, showing the main components and their specific interactions.

As shown in Figure 3.2 the LEGO-Mindstorms NXT brick acts as the central control unit of the device. During the measurement the strain and the force or resistance are evaluated synchronously by ATMEL-ATTiny261 microcontrollers. The data obtained by those "intelligent sensors" is sent on request to the NXT, which itself also acts as a communication device to the graphical LabView interface where the final interpretation, acquisition and recording of the measured data along with the measurement configuration and control is performed. To measure the tensile force the setup contains a force gauge (Vernier Software & Technology), which is fixed and stable mounted on the LEGO frame, with its sensor post being mechanically conjoined via an aluminium rod to one sample-clamp. The second clamp is mounted on a sliding carriage, which is mechanically connected to the linear actuators and a digital calliper (Wisent), allowing for uniaxial stretching of the sample and the precise measurement of the clamp displacement. The two linear actuators are driven by two Lego NXT 2.0 DC-motors with subsequent gearing.

The following sections give a detailed overview of the individual components, their specific tasks and the different data acquisition mechanisms.

3.1 Lego Mindstorms

Using LEGO for the build-up of a device which is supposed to conduct trustworthy scientific measurements may sound ludicrous on first thought. Impossible, some would even say. Almost everyone to whom I explained my intention of building an actual scientific measurement device with a kids-toy, would react either with rolling eyes, disbelieving stares, scepticism or even absolute denial. And as I knew LEGO from my childhood very well, when the idea came up to use LEGO as framework for a measurement device I could also be counted to the group of sceptics. At first. After challenging several drawbacks concerning mechanical stability, and data perception I realized what this "toy" is actually capable of. In combination with LEGO-Mindstorms it is a perfect tool for rapid-prototyping and even suitable for performing sophisticated measurements. This, the story behind LEGO and how the LEGO-Mindstorms system evolved over

time from being a simple toy to a highly adaptable data acquisition device
turned my initial notion by 180 degrees.

3.1.1 A Short History of LEGO Mindstorms

The LEGO NXT 2.0 package used for the build up of this device was
officially released in 2009 and is the latest of LEGO's Mindstorms robotics design kits[27]. This kit is the result of a long way of development and
the strategy to motivate children and students to discover science, technology, engineering, mathematics and robotics in a fun, playful and engaging
hands-on way [33].

The LEGO-Group, founded 1932 by the Danish carpenter Ole Kirk
Kristiansen, started the production of its construction toys in the late
1940s always under the restriction of backwards compatibility and inter-
operation ability with the "old" building blocks. These feature along with
the nowadays huge availableness of parts is one of the reasons why LEGO
was and still is an attractive option for construction and, therefore, enjoys
popularity throughout all age groups [34].

In 1980 LEGO started an initiative to implement smart toys for educational purposes in founding the Educational Products Department and a
collaboration with the Massachusetts Institute of Technology (MIT) Media Lab (more precisely Dr. Seymour Papert and the group of Dr. Mitchel
Resnick in 1985). Both shared with LEGO the view that a more promising way to educate children and students is by offering them hands-on
education (constructivism) rather than forcing them to memorize facts or
methodologies (instructivism). Resnick argued that *"training in computer
programming may be the most promising way to teach children about the
nature of problem solving"* and the best way to do this is to *"develop
new computational tools and toys that help people, particularly children, to*

---

[27]The next generation of LEGO's Mindstorms series, the LEGO Mindstorms EV3 will be released
in fall of 2013.

learn new things in new ways" [35].

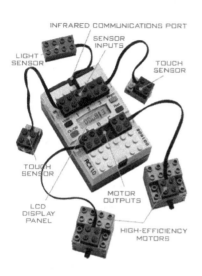

INFRARED COMMUNICATIONS PORT

SENSOR
INPUTS

LIGHT
SENSOR

TOUCH
SENSOR

TOUCH
SENSOR

MOTOR
OUTPUTS

LCD
DISPLAY
PANEL

HIGH-EFFICIENCY
MOTORS

Figure 3.3 – The Robotic Command eXplorer (RCX) programmable brick along with the sensor and actuators shipped with the LEGO Mindstorms Robotics Invention Kit.

Inspired by those intends Resnick's Epistemology and Learning group studied computer-based learning equipment, in particular self-programmable bricks with sensor and actuator interfaces and easy-to-use programming environment. Students should combine their practical use directly with engineering and technology and more importantly reckon them as educating games [34, 35]. This philosophy led to the first attempt to combine LEGO with modern computer technology, the LEGO-Technic Computer Control Series, released in 1985 and 1995 respectively. However the devices could not break into the market, since they were too expensive, too inflexible and, furthermore, public computer technology was still in the very early stages of its development. Nevertheless, the close collaboration of LEGO and the MIT resulted in the "red" and the "grey" programmable brick, which may be pronounced as the predecessors of the LEGO-Mindstorms system, which was finally released in 1998 as the LEGO-Mindstorms Robotic Invention Kit. This kit, including basic sensing and actuation hardware along with the RCX, an embedded microcontroller shown in Figure 3.3 and RCX-Code a simple graphical programming language became an instant commercial success, even called a "New Revolution" by the New York Times [34, 35].

Figure 3.4 – The NXT programmable brick along with the sensor and actuators shipped with the LEGO Mindstorms NXT 2.0 Kit.

Followed by this success, the second generation of LEGO-Mindstorms called the NXT System and the NXT 2.0 (Figure 3.4) were released in 2006 and 2009 respectively, offering state of the art hardware, newly designed sensors and actuators along with a new intuitive graphical programming interface. LEGO ships the package with its own graphical programming language called NXT-G, which is based on National Instruments LabView 7.1, being similar to RCX-Code and ROBOLAB, offering very basic control blocks for designing aforesaid robots. [33, 36].

3.1.2 The Hands-On Experience

Still finding its primary field in the playful approach of homebuilt, intelligent, independent, automatically roving robots, LEGO-Mindstorms (as intended) made its way into classrooms and project-based labs. Being highly adaptable, mostly by 3rd party improvements its nowadays mainly used for introductory computer courses, teaching of engineering, core-electronics, conception of hard- and software, maths, physics and robotics e.g. [37–42]. Nevertheless, due to its hands-on project character it also assists children in acquiring social skills like communication, teamwork, management and presentation [39, 40, 42–44]. Furthermore, LEGO also enjoys great popularity in scientific research. For instance the conception of prototypes for fully implemented industrial products [44], automated preparation of artificial bones [45], as platform for optical setups [46–51], microfluidic experiments [52], for astronomical purposes [53] or the use as simple prototyping platform being just some examples of educational and scientific purposes LEGO is used nowadays. Recently it also also found its way into space [54].

Nevertheless, although it is already widely used in scientific research, it is mostly used passively and for rather simple tasks. Therefore, and maybe as a trigger this work shows, that also trustable scientific measurements can be conducted with LEGO-Mindstorms and that it can be a powerful and cost-effective tool in research or rapid prototyping.

3.1.3 Not just a Kids Toy

LEGO-Mindstorms was initially intended to serve the age group of $10 - 14$ [35, p. 28], but since the introduction of the Robotic Invention Kit the Mindstorms systems inspired and motivated people of all age classes to put their own ideas into practice. Against expectation a huge community of hobbyists and professionals evolved around the toy, sharing numerous, partly highly sophisticated robot and machinery designs

Figure 3.5 – Fanbased NXT project *Cubestormer II* designed and constructed by Mike Dobson and David Gilday, currently holding the Guinness world record for "Fastest robot to solve a Rubik's Cube" [55, 56].

and ideas [36, 57–60], hardware improvements and extensions [61–65], 3rd party firmware and a vast number of programming language replacements and custom robot control software, including standard industrial languages and environments like C, C++, Python, MATLAB, LabView, Java and many more e.g. [60, 66–72]. With those extensions, the initial functionality and usability of the LEGO-Mindstorms system became widely extended,

offering the possibilities, software-tools and flexibility along with the processing power of the NXT for the construction of highly sophisticated
designs like the project described in this work or remarkable works like
the world-record holder for robotic Rubik's Cube solving by Mike Dobson
and David Gilday shown in Figure 3.5.

3.1.4 NXT Implementation in NI LabView

National Instruments released a special LEGO-Mindstorms NXT Toolkit
for its LabView software [66], which allows to bypass the limitations of
NXT-G, offering almost full LabView functionality for creating highly advanced control blocks, with the actual program running independently on
the NXT. In addition the NXT-plugin for LabView also offers a *"direct
control mode"* allowing to use full LabView functionality, because the actual program is running on a PC. This requires of course a permanent
connection of NXT and PC, either via USB or Bluetooth, but has the advantage of being able to realise a real-time control and data perception by
performing all computations on the PC and just sending move and sense
commands to the NXT. It offers the highest degree of flexibility and com-
puting power along with the advantage of having all time full control over
the device along with a graphical user interface. Thus this mode was the
chosen as control instance for the present stretching device. The LabView
user interface is shown in Figure 3.6.

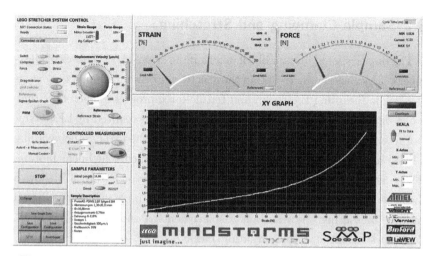

Figure 3.6 – LabView interactive user-Interface for control, data perception and evaluation for stress-strain measurements. The interface for resistance-strain measurements is fairly similar and is therefore not shown explicitly. The only significant difference is the additional option for selecting the desired measurement range.

Since all control mechanisms and the final data evaluation is done in LabView the NXT's main tasks remain to be controlling the drive of the motors, the handling of the communication, the data perception from the connected sensor board and the permanent delivery of the obtained data to LabView. So in principle the NXT acts as a communication bridge between the user interface and the actual measurement- and actuation-hardware. Thus, the NXT is in constant communication with the PC, whereby the user always stays in full control of the device. This permanent data flow permits adjusting all measurement parameters at run-time and allows for real-time tracking of the obtained data via two analogue displays and furthermore real-time graphical visualization of stress-strain or resistance-strain data. These recordings can be gathered either by pure manual control of the measurement procedures or by executing predefined measurement sequences. For further analysis those data recordings along with the measurement conditions are saved in tab-separated data files and, therefore, can be easily imported in standard data-evaluation software. In the present case subsequent data fitting and evaluation was done in Mathematica 9.0 and Origin 8.6.

3.2 Sample Clamping and Stretching

The device should be capable of determining stress-strain curves of elastomers. The easiest way to obtain those is by performing an uniaxial extension while recording the strain and tensile-force progress. Therefore, only
a one-dimensional movement is necessary. This was realized by LEGO linear actuators, which convert the rotary motion of the two parallel working
LEGO motors and the following gearing into a concurrent linear motion.
Those actuators shown in the upper right corner of Figure 3.7 allow a
displacement of 30mm and are conjoined to the sample grip and a digital
calliper measuring the resultant displacement as shown in Figure 3.15.
A basic schematic of the actuation mechanism is shown in Figure 3.7.
To prevent the linear actuators from reaching their drive-limitations two
limit-switches are placed on the frame as shown in Figure 3.8, immediately
stopping the motors when triggered.

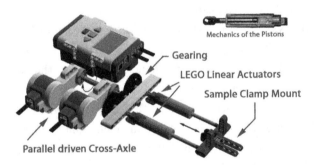

Figure 3.7 – Schematic, drawn in LEGO-Designer® showing the build-up of the actuation mechanism consisting of the drives, gearing, linear actuators and the mounting
for the sample grip[28].

---

[28]Piston sketch taken from [73].

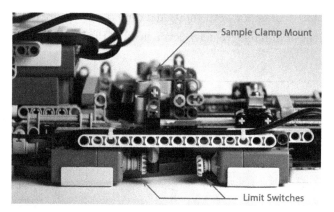

Figure 3.8 – End Switches disabling the motor drive when triggered by a post mounted on the sample-sledge.

In order to get comparable and reproducible results, the linear motion has to be performed with a well-known velocity. The used LEGO motors are DC-motors driven by a H-bridge driven pulse width modulation (PWM), embedded in the NXT. Inbuilt and external gearing increases the output torque. The motor-control LabView VI allows to configure the speed of those motors in the range of 0-100, so only the "power" rather than the "speed" may be chosen. With increasing load the revolution speed of the motor decreases, being typically for these kind of motors, resulting in a decreasing revolution speed with rising load. In addition those motors show different speeds in their rotational directions, resulting in different velocities for stretch and relaxation of the specimen. To lower at least the force dependence effects two motors are used in parallel, thus the output torque of the drive doubles theoretically.

However to completely eliminate both effects in order to guarantee a constant stretching velocity, a force-dependent motor power adjustment is worthwhile, whereas a simple correction function led to sufficient accuracy. Therefore, the stretching velocity dependent on the stretching force at a constant motor power was measured. This leads to a function $p(f, v, d)$ which gives the required motor power p for an applied load force f, velocity v and direction d. This ensures a constant stretching velocity over the whole force range for both rotational directions. Figure 3.9 shows the measurement of different velocity settings after the correction procedure.

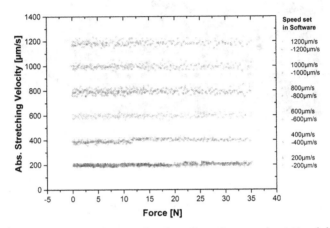

Figure 3.9 – Stretching velocities after force-dependence- and rotational-direction correction for different velocity settings. Positive velocities indicate a stretching, negative velocities indicate a relaxing movement of the stretching device.

The tight clamping of the specimen, is a crucial and maybe also a controversial part of the device. Although it was intended to use only LEGO parts for the device's framework, it remained impossible to guarantee a proper clamping of the samples. It was simply not possible to mount the elastomer samples without slippage. Therefore, a metal contraption had to be used. However, in order to completely avoid sample slippage double-faced adhesive tape has to be attached to the grips to increase the friction between them and the specimen. Nevertheless, the assembly[29] is rather simple as shown in Figure 3.10(b). Figure 3.10(a) shows the clamping contraption embedded in the LEGO frame.

---

[29]A detailed CAD is attached in the Appendix section.

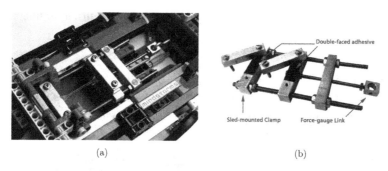

(a)   (b)

Figure 3.10 – a) Sample grip sledge embedded in the LEGO-device. b) Detailed close-up of the sledge.

Basically it consists of two thread rods, the sample grips and a linking bar to ensure directional stability during movement. One clamp is directly conjoined to the hook of the force gauge and, therefore, remains fixed during stretching of the specimen. The other clamp is steadily mounted onto the actuated sledge. Furthermore this solution offers the possibility for a well-defined tightening torque. This is important since elastomers are almost incompressible, and therefore are squeezed by the clamping to a certain extent. This leads to a significant change in sample length due to the resulting Poisson-stress. The effect increases with higher tightening torque. Therefore, in order to guarantee consistent measurements a well-defined tightening torque is essential.

3.3 Data Acqusition

Since the device is designed to measure two immediately connected properties (stress and strain or resistance and strain), it is important to acquire stress as well as resistance synchronously to the strain. The initial idea for the data acquisition was to connect for instance the force and strain sensor directly to the NXT and to read them separately using LabView commands. However, the LabView system in combination with the NXT cannot read the sensors at exactly the same time, leading to inevitable time-shifts between the two correspondent measurands. Therefore, data

points would not have exactly matching time-stamps, resulting in inherent
and irreproducible errors.

Due to this fact the decision fell for external data acquisition, for both
the displacement evaluation and the correspondent second property, force
or resistance. This was realized with a specially designed microcontroller
board shown in Figure 3.11.

Figure 3.11 – Data acquisition board containing two Atmel microcontrollers and
sensor readout electronics for acquiring force and resistance synchronously to the
strain information and for transferring the data to the NXT.

ATMEL-ATTiny261 controllers [74] and associated readout electronics[30]
ensure the acquisition of both properties at the same moment and therefore consistent data. The microcontrollers are in constant communication
with the NXT via a serial I²C interface, whereupon the acquired data is
transferred on request to the PC, bridged by the NXT. In the following
sections the data acquisition mechanisms will be described in detail.

3.3.1 Communication: Sensor Board - NXT

The possibility to connect I²C slaves to every sensor port is one of the most
powerful capabilities of the NXT. It allows for an almost endless expansion

---

[30]See Appendix section for the schematics and firmware.

of the Mindstorms system with sensors and actuators. However, in order to employ the sensor or actuator of choice, proper evaluation and control electronics, respectively, in combination with the necessary intelligence to communicate with the NXT via I^2C may be required. As shown in Figure 3.2 the I^2C interface allows the communication of the NXT with two AT-Tiny261 microcontrollers, responsible for the whole sensor-data perception procedures.

I^2C is a 8-bit oriented, two-wire bidirectional serial interface developed by Phillips in the 1980s [75], designed as master-slave bus system. Every communication between two participants is initiated and controlled by the master, the NXT. Theoretically the bus architecture allows for a total number of 127 slaves in one I^2C system. Nevertheless, communication is only possible between two devices at a time. However, the NXT does not fulfill the I^2C specification in all matters. Its output drivers are not capable of driving enough current, wherefore the bus only works properly by using $82k\Omega$ pull-up resistors [76, p. 9]. This allows to connect up to eight slaves rather than 127 on every sensor port line [65, p. 223]. Figure 3.12(b) shows the pin-assignments standard NXT-sensor cable.

In order to protect the I/O stages of the bus participants from failure due to electrical distortions, 150Ω resistors have to be placed in series to every slave on the SCL and SDA line, as schematically shown in Figure 3.12(a). Since master and slave are fed by the same power source (no galvanic isolation) an additional connection of the interface GND is not necessary.

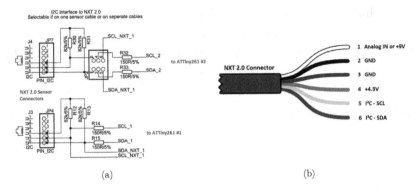

(a)   (b)

Figure 3.12 – a) Schematic circuit diagram of the connection of the NXT acting as I²C master and the ATTiny261 I²C slave devices. Note the 82kΩ pull-up resistors and the 150Ω series resistors on SDA (serial data) and SCL (serial clock) preventing the device from damage due to electronic distortions. b) Terminal pin assignment of the NXT sensor cable, whereas pins 5 and 6 are the I²C clock and I²C data lines respectively[31].

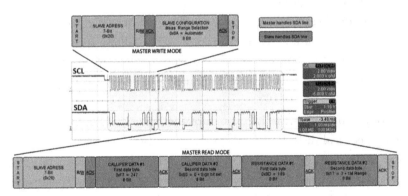

Figure 3.13 – I²C data transfer package showing the communication between the NXT (I²C master) and the resistance-strain sensor (slave). The waveforms were acquired with a LeCroy waveRunner 6100a digital oscilloscope.

As an example of a communication used in the measurement device Figure 3.13 shows a typical I²C package from the resistance-strain sensor. Since write and read requests cannot be carried out in one stream, the communication is split in two separate blocks. First the NXT, which in all cases acts as master, initiates the communication with a *start-condition*

---

[31]Sketch taken from [77].

followed by the unique 7-bit slave address and the low $R/\overline{W}$ bit, indicating that the master wants to write data to the slave. The addressed slave acknowledges this by subsequently forcing the SDA line low (ACK). The following data byte tells the slave in which measurement range the next resistance measurement should be carried out, whereby the slave again acknowledges the correct reception. The communication block is terminated by a *stop-condition* releasing the SDA and SCL lines for the next communication.

After this the master wants to receive the last resistance-strain data point. Therefore, it initiates a new communication with a start condition and the same slaves address, but this time with a data request indicated by the set $R/\overline{W}$ bit. After acknowledgement by the slave it immediately puts the data byte-wise on the SDA line with the master controlling the SCL and acknowledging every byte until it quits the transfer with $NACK$ and a stop condition.

The I^2C communication between the NXT and the stress-strain sensor is equal to the one shown above. The only differences are, that it has a different identification address, no data is written to the slave, since the sensor needs no configuration and the last two read bytes contain the prevailing tensile force. The complete realisation of hard- and software is shown in the Appendix Section.

3.3.2 Strain-Measurement

As further discussed in Section 4.1.2, to measure the strain of a sample simply the displacement of the sample grips has to be acquired. In principle two methods have to be taken into consideration, direct and indirect measurements. An example for an indirect measurement would be to apply a rotary encoder to the motor axis and counting the increments during the movement. With the number of increments the exact driving angle can be determined and therefore, with knowledge of the transmission ratio of the applied gearbox and the linear actuators, the linear displacement can be calculated. Since the LEGO motors are shipped with included rotary encoders with a readout accuracy of 1° this would be the most cost-effective option and not require external hardware of any sort. However, there is one huge drawback. To achieve an accurate measurement the transmission and linear actuation are not allowed to have any mechanical clearance. Therefore, high precision transduction and transmission units would be necessary, requirements which the LEGO gear-wheel transmission and the linear transducers certainly do not satisfy as shown in Figure 3.14. In this measurement the error of the displacement was determined by comparison with a manual measurement via a digital calliper. It is clear, that the determination of the displacement by evaluation of the internal rotary encoders is not suitable in order to guarantee an acceptable accuracy, since errors in the order of 1mm would be rather guessing than measuring.

Therefore, a direct measurement method is worthwhile and is realized by mounting a strain sensor directly onto the moving sledge. The maximum displacement is limited by the LEGO linear actuators to an absolute displacement range of 30mm. In this particular range the requirements for the applied strain sensor are a high measurement accuracy, an easily implementable interface, unproblematic mounting on the LEGO framework and rather low costs.

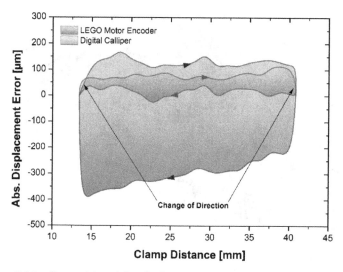

Figure 3.14 – Comparision of the absolute measurement accuracy of indirect measurement via the embedded rotary encoders and the direct measurement via the mounted digital calliper. The linear actuators were fully expanded and retracted.

There are several displacement gauges on the market being compatible with the LEGO Mindstorms system, like the ultrasound and light detection sensors included in the NXT 2.0 set [33] or electro-optical [62] sensors. However, all of them are appropriate at most for proximity detection, but not for the precise determination of linear displacements. Therefore, a self-built solution was inevitable. Potential candidates for this task range from resistive strain gauges to capacitive or inductive gauges, but also techniques which apply optical or acoustic effects. However, most of these sensor applications require rather complex and expensive evaluation electronics. Therefore, to fulfil the mentioned requirements and to reduce complexity, a fully implemented measurement solution was favourable. The decision fell for a rather exceptional approach shown in Figure 3.15.

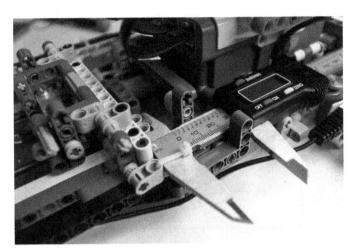

Figure 3.15 – The digital calliper installed on the stretcher. The top is affixed on the sample sledge while the body remains stationary on the LEGO frame, therefore, measuring the linear displacement of the sample grip relatively to the stationary frame.

An easy and cheap way to achieve a fairly accurate displacement measurement is to use a cheap no-name digital calliper, available in every hardware store. They often can be purchased for less than 20 Euro, offer an absolute accuracy of $10\mu m$ and mostly rely on a similar hard- and software design throughout the distributors. With the application of such a device, the measurement error decreased by almost a factor of 6, compared to the rotary encoder solution as shown in Figure 3.14, since any mechanical clearance issues are almost neglectable. Nevertheless, due to inevitable mechanical instabilities and the fact that the calliper mounting causes a certain bending moment on the specimen sledge, which accounts for the slip at the turning points, still an error around $100\mu m$ has to be considered.

The electronics underneath the different distributors branding is often identical. The important part however is, that most of those callipers offer a very simple digital interface. The difference to name-brand callipers, which often provide rather sophisticated serial communication [78], is that this feature is often not mentioned, the interface is rather simple and suitable interface cables are either expensive or cannot be purchased at all.

For the present design a digital calliper (Wisent) was used, which offers such a simple serial interface. Custom but yet simple readout electronics was designed, which will be described in detail in the following section.

3.3.2.1 Interfacing the Calliper

The digital calliper interface shown in Figure 3.16(a) is arranged as a 4-pin connector consisting of the voltage supply lines and the two-wire serial interface, clock- and data.

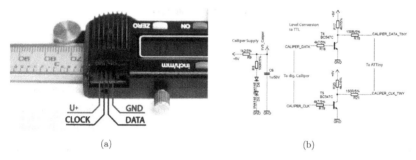

(a)   (b)

Figure 3.16 – a) Digital interface of the used digital calliper (Wisent) consisting of the voltage supply and the two-wire serial interface. b) Circuit diagram of the calliper voltage supply and the level conversion to TTL via open-collector circuit.

It is important to note that regardless of the calliper being switched on or off it keeps providing the data stream, which leads to a constant current draw from its supply due to the level conversion to TTL via the open collector circuit shown in Figure 3.16(b). This conversion is necessary to allow the microcontroller to pick up the signals. However, this interface connection would lead to a fast discharge of the internal battery. Furthermore, it was found, that the calliper data-rate strongly depends on its supply voltage, as shown in Figure 3.17. Therefore, it is advisable to provide a stable external power supply.

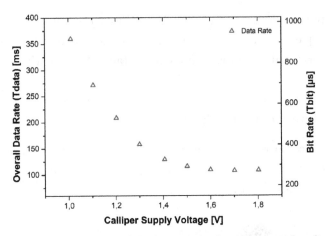

Figure 3.17 – Dependence of the data rate on the calliper supply voltage.

In the current design a supply voltage of 1.7V was used in order to guarantee a fast data acquisition without the risk of damaging the internal electronics. With that every T_{DATA} the calliper sends a 24-bit burst consisting of six 4-bit nibbles, representing the momentary displacement value x100 plus the sign bit in the upper nibble. The data is sent as least significant bit (LSB) -first, whereupon every data bit being valid at the falling edge of the clock line as shown in Figure 3.18.

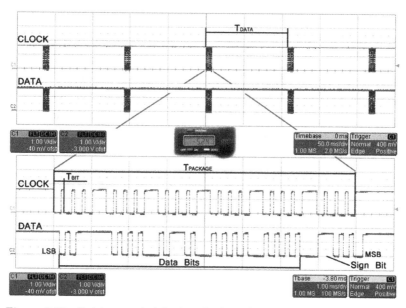

Figure 3.18 – Serial protocol of the digital caliper showing the Clock- and Data line at a vernier displacement of -5.24mm and a supply voltage of 1.7V. The traces were measured with a Wave Runner 6100a digital oscilloscope (LeCroy).

$$\text{DATA Stream (LSB first)} : \underbrace{\underbrace{0011\ 0000\ 0100\ 0000\ 0000}_{524_{\text{DEC}}}\ \underbrace{1000}_{\text{Sign bit set}}}_{\text{Displacement value: } -5.24\text{mm}} \qquad (3.1)$$

Table 3.1 – Digital calliper timings at a supply voltage of 1.7V.

	T_{DATA} [ms]	T_{PACKAGE} [ms]	T_{BIT} [μs]
Duration	107.5	7.9	275.6

The way of choice to read the calliper value is, therefore, to assign the clock line to an external interrupt of the microcontroller and constantly shift the corresponding data bit into a shift register until the complete 24-bit burst package was read. Since the actual value is delivered, no further computation besides handling the sign-bit is needed. Due to the rather low calliper data-rate of ca. 100ms per displacement value the overall perception rate is also limited to ca. $10\frac{\text{Data Points}}{\text{s}}$.

The calliper is not capable of measuring its absolute displacement, but needs to be referenced on every power up of the device. Therefore, prior to each measurement the initial sample length l_0 has to be determined by manual measurement of the samples dimensions at 0N tensile force. Henceforward, the sample-strain λ is calculated

$$\lambda = \frac{l}{l_0} \tag{3.2}$$

with the ongoing length l of the sample specimen.

3.3.3 Stress-Measurement

To measure the nominal mechanical uniaxial stress σ_N, according to Section 2 it is necessary to know the tensile force f along with the initial cross-section A. The cross-section can be determined by measuring the width b and the thickness d prior to clamping the sample with a digital calliper and a micrometer gauge respectively, resulting in

$$A_0 = bd \tag{3.3}$$

$$\sigma_N = \frac{f}{A} \tag{3.4}$$

3.3.3.1 Measurement Method - May the force be with us

For the determination of the tensile force f a dual-range force gauge (Vernier) [61], offering a fully implemented, mechanically and electronically LEGO-compatible solution is used. The sensor contains standard resistive strain-gauges in half-bridge configuration mounted on a bendable beam, generating a voltage proportional to the applied force. In addition the sensor voltage is internally conditioned whereby it is possible to manually change the measurement range to either $\pm 10N$ or $\pm 50N$, with absolute accuracies of 0.01N and 0.05N respectively. In both cases the sensor delivers a stabilized $0 - 5V$ output voltage. This voltage is picked up on the one hand directly by the NXT for stretching speed regulation[32] and on the other hand by a 10-bit analogue-digital converter (ADC) embedded in the microcontroller for synchronous acquisition of the tensile force to the strain data.

The sensor is steadily mounted on the LEGO frame. Its sensing post is conjoined with one sample grip via an aluminium rod, shown in Figure 3.19, allowing for gauging the tensile force on the sample upon deformation.

Figure 3.19 – Dual-Range force gauge (Vernier) mounted and fixed in the setup, while being mechanically conjoined to the clamping contraption.

Since there is no calibration information available from the manufacturer, the exact characteristics of the used force sensor needed to be analysed. Therefore a calibrated FMI-220 force gauge (Alluris) in combination with an U1273A digital multimeter (Agilent) was used to determine the corresponding fit-parameters for offset and gain correction. The results are shown in Figure 3.20.

---

[32]See Section 3.2

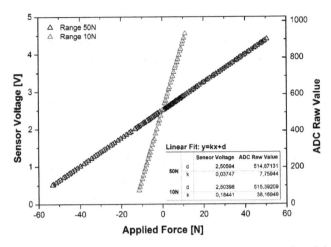

Figure 3.20 – Calibration data concerning sensor voltage and raw-value of the 10-bit ADC for the ±10N and ±50N measurement range of the used dual-range force gauge.

In both cases the output signal of the force gauge is digitalized by a 10-bit ADC embedded in the controller of the NXT and the external microcontroller on the sensor board (see Figure 3.2) respectively. According to the data-fits in the latter figure the full conversion range of $5V$ yields a significantly larger measurement range than given in the manual, resulting in an overall resolution of

$$Res \ [\text{N/Bit}] = \frac{\text{Range [N]}}{2^{10}\text{Bit}} \tag{3.5}$$

Table 3.2 – Theoretical measurement range and measurement resolution for tensile force according to the full voltage conversion range of $0 - 5V$.

Range [N]	f_{0V} [N]	f_{5V} [N]	Resolution [N/Bit]
±50	-66.97	66.47	0.130
±10	-13.58	13.54	0.027

To reduce errors caused by ADC quantization and jitter (typically ±1 LSB [74]), signal noise, and other parasitic effects, its advantageous not to just take one force value, but to repeatedly sample the signal and take the arithmetic mean. Those samples are taken on the first calliper-clocks which trigger the external interrupt. Since an AD-conversion needs 13 ADC-clock cycles [74], a full conversion needs 104μs, taking into account that the ADC-clock equals 125kHz. This ensures, considering the calliper timings shown in Table 3.1, that on every external interrupt the previous conversion has finished. In the present design a sampling number of 4, resulting in an overall conversion time of $\backsim$ 1ms led to satisfying results. Nevertheless, according to the datasheet an ADC inaccuracy of $\pm$1 LSB has still to be taken into account.

Furthermore, in order to guarantee accurate results, the stress-strain curve of a measurement where no specimen is attached, should be ideally a 0-line over the whole strain range. Since factors like friction, lack of mechanical stability, vibration effects and ADC jitter cannot be fully eliminated, a certain error is unavoidable. Figure 3.21 shows a typical 0-curve, measured from limit-switch to limit-switch with a velocity of 500$\frac{\mu m}{s}$.

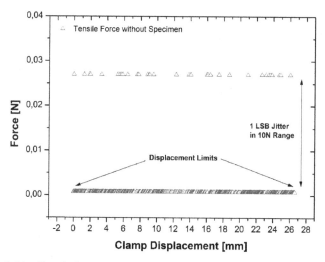

Figure 3.21 – Tensile force while driving three times from end-switch to end-switch with no sample attached, measured in the $\pm$10N range with a velocity of 500$\frac{\mu m}{s}$.

3.3.4 Resistance-Measurement

In addition to the stress-strain measurement also a resistance-strain measurement was included in the setup, with the intention to allow having a
cheap, but, nevertheless, accurate possibility to characterize for instance
the varying resistance of stretchable electrodes when stretched as reported
in e.g. [79–82].

So the question that remains is, if it is reasonable to employ a self-
contained solution like using a standard multimeter and interfacing it via
LabView or designing a custom-made solution. The choice fell for the latter, inasmuch being cheap, simple and flexible. Moreover it embeds the
concept of the "Smart-Mindstorms-Sensor" concept resulting, in combination with the stress-strain measurement, in a full self-contained solution.
Furthermore, with this approach a trustworthy synchronous data acquisition of strain and resistance data is guaranteed.

3.3.4.1 Measurement Method

In principle the measurement of resistance mostly relies on the measurement of a voltage drop across the test specimen under controlled conditions. By passing a well defined current I_R through the specimen, resulting
in a voltage drop U_R, the sample resistance R is calculated by Ohm's Law.

$$R = \frac{U_R}{I_R} \tag{3.6}$$

Since the resistance measurement is no aftermarket sensing solution like
the force gauge, special readout electronics were designed and are described
briefly. An abridgment from the circuit diagram of the present design is
shown in Figure 3.22. The full schematic is attached in the Appendix
Section.

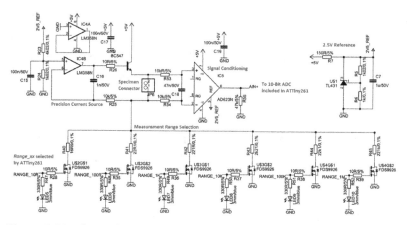

Figure 3.22 – Circuit diagram of the resistance measurement, showing the precision constant current source, the signal conditioning via instrumentation amplifier and the measurement range selection mechanism.

The precision constant current source consists of a 2.5V reference voltage U_{REF} generated via a TLV431 reference diode (Texas Instruments) [83], which also acts as the reference for the AD-conversion. This reference voltage is used to create a 0.5V reference $U_{0.5V}$ via a precision voltage divider, which is then fed into a general purpose LM358 operational amplifier (Texas Instruments) [84]. This OpAmp regulates the voltage drop over the reference resistor by controlling the base current of a BC547 driver transistor (Fairchild) [85] resulting in a precise constant current I_R through the specimen connected to the pins JP6.

$$U_{REF} = \frac{1.24V \cdot (R_5 + R_6)}{R_6} = 2.505V \tag{3.7}$$

$$U_{0.5V} = \frac{U_{2.5V} \cdot R_{24}}{R_{23} + R_{24}} = 0.507V \tag{3.8}$$

$$I_R = \frac{U_{0.5V}}{R_{REF} + R_{DS_{ON}}} \tag{3.9}$$

This mechanism works, as long as the LM358 is able to deliver enough output voltage in order to ensure the 0.5V over the reference resistor. If this is no longer possible, for instance if the attached specimen has a too high resistance or no specimen is connected (open-line-condition), the LM358 output saturates. This behaviour would result in a drastic limitation in either the measurement precision or the measurement range. By employing selectable reference resistors, different measurement ranges are possible, allowing for higher precision throughout the desired measurement range of $0 - 1M\Omega$. These resistors are selectively activated through n-channel MOSFETs. What is fairly important in this matter, is the use of MOSFETs with a low drain-source on-resistance ($R_{DS_{ON}}$) in order to minimize measurement errors. For the current design a FDS9926 dual n-channel FETs (Fairchild) [86] with an on-resistance of $R_{DS_{ON}} \leq 30m\Omega$ is used.

The following table shows the measurement conditions for the different measurement ranges. In order to avoid damaging the specimen by too high currents, the measurement current was limited to 30mA in the lowest measurement range, wherefore an additional amplification by 8 prior to the AD-conversion was necessary to extend the signal to the full conversion range. In the control software the final measurement range can be selected either manually (manual) or by the device itself (auto).

Table 3.3 – Conditions of the resistance measurement for the different measurement ranges with R_{REF} being the reference resistance, I_{REF} the measurement current and gain the factor by which the U_S is amplified prior to the AD-conversion.

Range	R_{REF} [Ω]	I_{REF} [A]	Gain	Resolution [Ω/Bit]
10Ω	16R9	30.0m	x8	0.01
100Ω	22R1	22.9m	x1	0.11
1kΩ	221R	2.29m	x1	1.07
10kΩ	2k21	229μ	x1	10.7
100kΩ	22k1	22.9μ	x1	106.6
1MΩ	221k	2.29μ	x1	1066.3

3.3.4.2 Signal Conditioning and Measurement

The regulated current leads to a precise voltage drop across the unknown resistor. This voltage is picked up, slightly low-pass filtered to reduce high frequency jitter and conditioned by an AD623 instrumentation amplifier (Analog Devices) [87]. This guarantees a well-defined low impedance measurement voltage U_S, which is afterwards fed into the ADC embedded in the microcontroller. Very important is to ensure that the input voltages of the AD623 do not exceed $U_+ - 1.5V$ and $U_- - 0.15V$ on the "+" and the "-" input pin respectively as stated in the datasheet. The parameters U_+ and U_- in this case being the positive and negative supply voltages of the AD623. Therefore, it is essential to place the BC547 driver transistor on the 5V supply line, assuring the necessary voltage drop to at least 3.5V. This enables for proper regulation and measurement. The effective "working-range" for the voltage drop on the specimen is therefore reduced to $3.5V - U_{0.5V} = 3V$.

The AD623 is single supplied and thus, although it is a rail-to-rail type, its output voltage cannot be lower than $0.01V$. This would not allow to measure resistances lower than 0.33Ω, taking the measurement current at the lowest measurement range into account. Therefore, an offset voltage of U_{REF} needs to be added to the differential input voltage U_R. This reference voltage also acts as reference 0-potential for the AD-conversion and is also assigned as ADC-reference voltage, defining the measurement range. This ensures a defined basis for the actual measurement and also a defined working range for the AD-conversion. The resolution for the measurement can be calculated by the following equations, with R_{MAX} being the maximum resistance to be measured in the selected range.

$$R_{MAX} = \frac{U_{REF}}{U_{0.5}} \cdot R_{REF} \qquad (3.10)$$

$$Res[\Omega/Bit] = \frac{R_{MAX}}{2^{10}\text{Bit}} \qquad (3.11)$$

In addition to the pure manual selection of a measurement range it is advantageous to allow an automatic configuration of the ideal measurement range. This ensures that the measurement is always carried out in the most accurate range. To realize this, the full range of the AD-converter (in the case of the used microcontroller this would be $2^{10} = 1024$) has to be segmented into three parts, the actual measurement region and in a lower and higher region as shown in Figures 3.23. A signal voltage approaching the borders is detected and the measurement range re-configured.

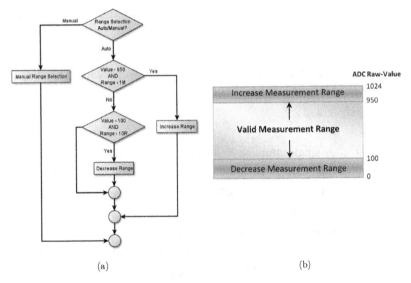

(a)   (b)

Figure 3.23 – Segmentation of the AD-converter range in order to allow the controller to check and change the resistance gauging range to ensure ideal measurement conditions.

If a signal exceeds one of the borders, the microcontroller instantly changes the measurement range by switching to another reference resistor, resulting in switching the measurement range according to Table 3.3. This mechanism ensures that the measurement is always carried out in the ideal range. The AD-conversion procedure is identical to the one already described in Section 3.3.3.

4 Pertinent Example Measurements

"The most exciting phrase to hear in science, the one that heralds new discoveries, is not 'Eureka!' (I found it!) but 'That's funny ...'"

— Isaac Asimov

Now that the necessary theory of rubber-elasticty and the measurement method along with the design of the measurement setup was laid out, a few example measurements are shown in the following chapter. In particular the Young's Modulus of polydimethylsiloxane (PDMS) with different degree of cross-linking and its dependence on the stretching velocity is determined with the methods explained in Section 2. The second part focusses on the electromechanical characterization of stretchable electrode films on PDMS.

4.1 Determination of Young's Modululs

4.1.1 Sample Preparation

The elastomer used for the tests, polydimethylsiloxane (PDMS) Sylgard 184 from Dow Corning, is a two component system consisting of a pre-polymer and a cross-linking agent which are mixed in weight ratios of 5:1, 10:1 and 20:1. With the amount of cross-linker the degree of cross-linking of the elastomer is controlled resulting in different viscosities and therefore different Young's moduli. A higher amount of cross-linker results in a stiffer elastomer. After weighing in polystyrene petri-dishes, the compounds are mixed for 5 minutes with a sterile spatula and afterwards degassed in an desiccator to remove air-bubbles. The dishes are then put to rest on a levelled slot in an oven for one hour to allow the mold to form a uniform film with 1mm thickness. After this levelling process the mold is cured in the oven for 15 hours at 65°. After curing the standardized dumb-bell shaped samples are cut out from the PDMS films using a laser-cutter (Trotec Speedy 300). This cutting process leaves residues at the cutting edges which are removed by washing in distilled water and ethanol. The samples are afterwards dried for 24h at room temperature.

4.1.2 Sample Geometry

To allow a trustworthy determination of the Young's modulus from a stress-strain measurement, the stress as well as the strain have to be uniform over the whole gauging range. To realise uniform stress and strain, a special sample geometry has to be used. The shape, the clamping and the methodology on how to determine stress and strain data from the sample are described in the ISO527-3 standard [88]. The standard defines a dumbbell-shaped specimen shape as shown in Figure 4.1.

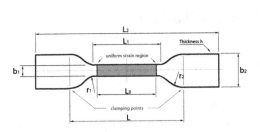

	Dimension [mm]
L_2	≥ 35
b_2	6 ± 0.5
L_1	12 ± 0.5
b_1	2 ± 0.1
r_1	3 ± 0.1
r_2	3 ± 0.1
L	20 ± 2
L_0	10 ± 0.2
h	≥ 1

Figure 4.1 – Dumbbell-shaped test specimen 5B according to ISO527-3

Table 4.1 – Associated dimensions.

This shape ensures a large bearing-area along with producing a homogeneous, uniform stress-strain region at the narrow bar (blue region in the latter figure). The geometry of this bar is rather unaffected by the sample grips, which squeeze the sample to a certain degree and, therefore, create a highly inhomogeneous region in their adjacency. The cross-section over the whole "uniform region", therefore, stays well-defined also for high strains allowing the assumption of pure uniaxial strain in this region. This allows for the use of the approximations presented in Section 2.6.

To determine the sample strain, only the strain of this uniform region is needed. Typically a separate pair of clamps is attached to the sample at both ends of the uniform strain region. The change in distance of those clamps in then used to calculate the strain. Another rather new method is to apply optical marks on the specimen and use camera or laser assisted tracking software [89]. A third method often used is to simply measure the displacement of the sample grips. This has the advantage that the sample itself is not affected by additional gauging measures. However, it is important to be aware of the significant influence of the dumbbell geometry on the total strain progression, while only the uniform region provides reliable strain information. Thus, certain correction measures are necessary. Schneider et al. showed by manual measurement and by finite element (FE) analysis of a ISO527-3/5A conform PDMS specimen[33] that simply a strain-correction factor $m = 2$ holding for strains $\leq 40\%$ is needed [90].

$$m = \frac{\Delta L}{\Delta L_0} \quad \rightarrow \quad \varepsilon_{uniform} = \frac{\varepsilon_{total}}{m} \tag{4.1}$$

To confirm these findings also for the ISO527-3/5B specimen, a similar measurement was conducted. The uniform region of interest was optically marked and its progression manually read for different degrees of strain, whereas every data point represents the mean of 3 independent measurements. For comparison also the strain according to the clamp displacement was determined. The reference length for both strains was $l_0 = 10$mm. The initial grip-grip distance needed to be readjusted to $L = 24.6$mm, since due to the clamping the (incompressible) sample is squeezed and thus pressed out of the grips to a certain extent. The strains can then be calculated with

$$\varepsilon_{total} = \frac{\Delta L}{l_0} \qquad \varepsilon_{uniform} = \frac{\Delta L_0}{l_0} \tag{4.2}$$

---

[33]Dow Corning Sylgard 184 with a prepolymer to cross-linker ratio of 10:1.

Figures 4.2(a) and (b) show the result of this measurement, which is in good agreement with the prediction and the experiment by Schneider et al. The strain corrected by a factor of $m = 2$ matches the strain of the uniform region in good accordance up to strains $\leq 60\%$ where slipping of the sample from the grips occurs. With increased tightening torque also higher strains are possible, but also increases the risk of rupture at the point the sample is clamped.

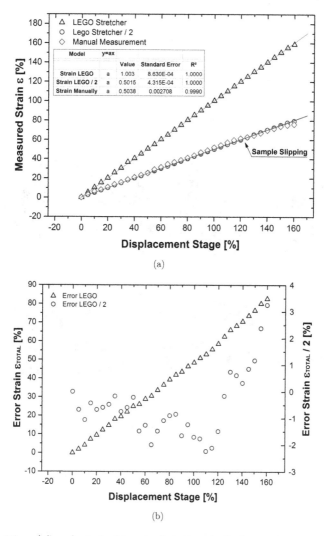

Figure 4.2 – a) Sample strain determined via the grip displacement ε_{total} compared to the manual measurement of the uniform strain region $\varepsilon_{uniform}$ b) Deviation of the measured strain to the strain in the uniform strain region.

In addition the homogeneousness of the strain in the uniform region was analyzed with an optical, contactless strain visualization system (Aramis, GOM) [91]. The measurements were conducted at the Institute of Polymer Product Engineering (IPPE) at the JKU Linz. The Aramis setup collects images prior to and during the deformation and compares the images to the reference image of the undeformed state. To allow for such strain studies the sample was coated with a noise pattern. The Aramis system subsequently tracks the movement of the pattern during deformation enabling 2D- or even 3D deformation studies. Figures 4.3(a) to (d) show the Aramis recordings for different degrees of strain.

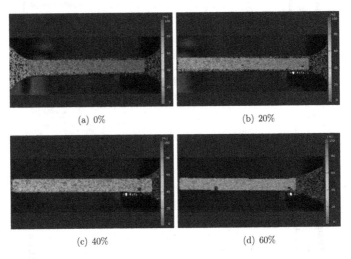

(a) 0%   (b) 20%

(c) 40%   (d) 60%

Figure 4.3 – ARAMIS 2D strain analysis showing the strain homogeniousness in the uniform strain region of a ISO527-3/5B compliant PDMS specimen (1:10 Dow Corning Sylgard 184).

The uniformity as well as a trustable strain information when measuring the grip displacement is guaranteed for elastomer specimen conforming to ISO527-3. Figures 4.4(a) and 4.4(b) show a sample used for the measurements in the next section. Nevertheless it is often not practical or even not possible to realize an ISO527-3 conform sample geometry. In such cases, like the stretchable electrodes in Section 4.2, also straight cuboid specimen may be used as stated in the ASTM standard [92]. To reduce the effect of the inhomogeneities at the clamping points, the sample has to be of sufficient length. The measurement procedure itself is similar to the dumbbell specimens. For the following stress-strain measurements the standardized dumbbell-shaped samples are used.

(a) (b)

Figure 4.4 – a) ISO527 conform PDMS sample b) clamped and stretched in the LEGO setup.

4.1.3 Comparison of Fit-Models

When determining the Young's Modulus it is important to know which model fits the acquired data best. Figures 4.5(a) and (b) show a typical, representative stress-strain recording of a 10:1 PDMS sample up to 80% strain after the previously mentioned ISO527-3 strain correction. The curves are substantially reversible, so the tested PDMS samples did not show significant viscous behaviour. All following plots show the nominal stress σ_N. The data was fitted with Mathematica 9.0.

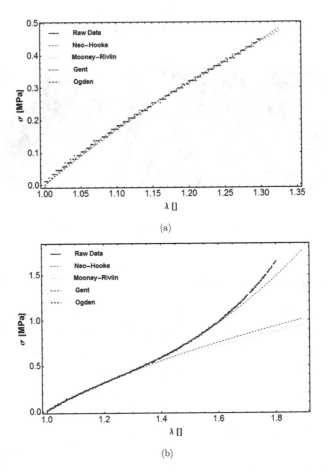

(a)

(b)

Figure 4.5 – a) Stress-strain recording of Sylgard 184 with a mix ratio of 10:1 and a stretching velocity of $200\frac{\mu m}{s}$. The data was fitted with the pre-mentioned elasticity models up to 30% strain. b) fully recorded data up to 80% strain with the same fit data.

The fitting in the latter plots was done only for the region of small extensions $\leq 30\%$. All considered models fit this (linear) region sufficiently as shown in Figure 4.5(a) and can, therefore, be applied for the determination of the Young's modulus according to Equations 2.91, 2.92, 2.93, 2.94 and 2.95 given in Section 2.6 when assuming a Poisson ratio of $\nu = 0.49$.

Table 4.2 – Prediction of shear modulus G and Young's modulus E of PDMS Sylgard 184 with a mix ratio of 10:1. The models were applied to the region of small extensions $\leq 30\%$. Data-fitting shown in Figure 4.5.

	Neo-Hooke	Mooney-Rivlin	Gent	3rd order Ogden
G [MPa]	0.63	0.65	0.63	0.73
E [MPa]	1.88	1.95	1.88	2.21
$\lambda_{Rupture}$ []	-	-	1426.9	-

The prediction of the Young's modulus E is in good accordance through-out the models, only Ogden yields some deviation from the other models. Nevertheless, as shown in Figure 4.5(b) the continuative prognosis of the behaviour for strains $\geq 30\%$ shows a rather poor agreement for all models. This is obvious, since the stress-strain becomes highly non-linear. For strains $\geq 40\%$ the elastomer enters the *strain-stiffening* region, resulting in an increased hardening of the material upon further stretching, leading to a steadily increasing slope of the stress-strain curve until the point of rupture. This point is predicted by the Gent-Model due to the discontinu-ity of the ln term. Nonetheless, to predict this point the dataset cannot stop in the linear region as apparent in Figure 4.5(b). At least the onset to the strain-stiffening region is necessary to allow a sufficient prognosis of the rupture point.

Figure 4.6 shows the same data, but this time the models were fitted for the whole data range up to 80% strain.

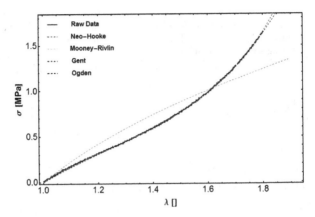

Figure 4.6 – Stress-strain recording of Sylgard 184 with a mix ratio of 10:1 and a stretch ratio of $200\frac{\mu m}{s}$. The data was fitted with the pre-mentioned elasticity models up to 80% strain.

As apparent from the latter figure, Neo-Hooke and the Mooney-Rivlin definitely reached their functional limits. As soon as the data leaves the initial linear region these models are no longer suitable. Furthermore, they predict almost the same trend as they tend to overlap for such data-sets. This similarity was investigated in many measurements and is also the case in this plot. The more sophisticated Gent and Ogden theories fit the whole highly non-linear data-set in very good agreement. Especially the Gent approach is rather prodigious since it only needs two fit-parameters compared to the 3rd order Ogden needing six. In addition Gent also yields the point of rupture to occur at 122% strain, which is in agreement with literature.

Table 4.3 – Prediction of shear modulus G and Young's modulus E of PDMS Sylgard 184 with a mix ratio of 10:1. The models were applied to the full measured data-set $\lambda \leq 80\%$. Data-Fitting shown in Figure 4.6.

	Neo-Hooke	Mooney-Rivlin	Gent	3rd order Ogden
G [MPa]	0.83	0.83	0.59	0.66
E [MPa]	2.49	2.49	1.75	1.99
$\lambda_{Rupture}$ []	-	-	1.22	-

The predictions of Gent and Ogden for the Young's modulus differ significantly and, furthermore, differ from the low-extension predictions shown in Table 4.6. In literature the Young's modulus for Sylgard 184 with a mixing ratio of 10:1 is mostly found to be in the range of 1.8MPa e.g. [90,93], which could be proofed with rather good agreement with both fitting options. A more precise prediction is found in the next paragraph.

The question that remains is, which prediction for the Young's modulus is the most suitable one. Generally speaking each of the mentioned models is suitable, when applied to a suitable data-set. In publications the use of Neo-Hooke and Mooney-Rivlin are mostly found e.g. [90,94–96]. However, the Gent model is suitable to fit also the non-linear region along with the prediction of the rupture-strain, while only needing two fit-parameters. This is the reason why this model is favoured in this work. All upcoming data-fits for stress-strain curves are based on the Gent model.

4.1.4 Young's Modulus for different degrees of cross-linking

As already mentioned the degree of cross-linking is controlled by the amount of cross-linking agent added to the pre-polymer. With increased number of cross-linkages per unit-volume the elastomer gets stiffer as already depicted in Section 2.3. In particular three different configurations with mix-ratios of 5:1, 10:1 and 20:1 were tested. Therefore, three samples were fabricated for each mix-ratio and analyzed. Figure 4.7 shows representative stress-strain recordings.

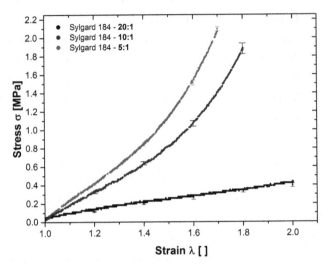

Figure 4.7 – Stress-strain recording of Sylgard 184 with mix ratios of 5:1, 10:1 and 20:1. The samples were tested with a stretch ratio of $200\frac{\mu m}{s}$. The data was fitted with the Gent model up to 70%, 80% and 100% strain respectively.

Table 4.4 – Prediction of the Young's moduli E and the rupture points of PDMS Sylgard 184 with mix ratios of 5:1, 10:1 and 20:1 when stretching with a strain-rate of $200\frac{\mu m}{s}$. The data was fit with the Gent model, applied to the full data-sets.

	5:1	10:1	20:1
E [MPa]	2.33 ± 0.06	1.83 ± 0.05	0.68 ± 0.04
$\lambda_{Rupture}$ []	2.04 ± 0.01	2.17 ± 0.01	-

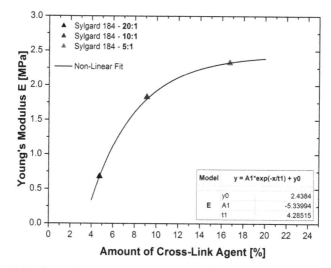

Figure 4.8 – Dependency of the Young's modulus on the mix-ratio based on the measurements from Figure 4.7 and Table 4.4

As shown in the latter plots, the Young's modulus is varied with the amount of cross-linking agent in the mold, whereby the dependence of the Young's modulus on the mix-ratio is found to be highly non-linear. Along with the stiffness of the material also the maximum strain $\lambda_{Rupture}$ decreases. The maximum strain of the 20:1 PDMS could not be determined, since the maximum displacement of the stretcher was reached before any strain-stiffening effects occurred. Khanafer et al. also investigated the effect of diffferent mixing-ratios for Sylgard 184, in particular molds with ratios from 6:1 to 10:1. They found, that the Young's modulus peaks at a ratio of 9:1 and then rapidly falls with increasing amount of cross-linking agent. For 6:1 the Young's modulus remained being only 53% of the 10:1 mixture [97]. This behaviour could not be reproduced.

4.1.5 Young's Modulus for different rates of strain

The Young's modulus of viscoelastic materials shows a strong dependence on the strain-rate during testing. Silicones like Sylgard 184 only show little viscoelasticity, depicted by the lack of a considerable hysteresis in the stress-strain behaviour. Nevertheless, an effect of different stretching velocities is observed. For the tests samples with a mixing-ratio of 10:1 were used. The sample was stretched with strain-velocities of 200, 600 and $1000\frac{\mu m}{s}$. The effect is indeed rather small, since the stress-strain curves for different stretching-velocities almost overlap as shown in Figure 4.9.

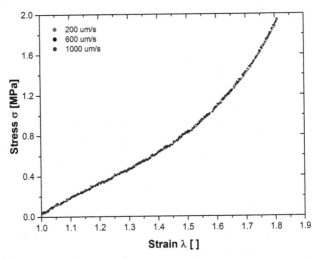

Figure 4.9 – Overlay of three representative stress-strain measurements of 10:1 Sylgard 184 for different stretching velocities.

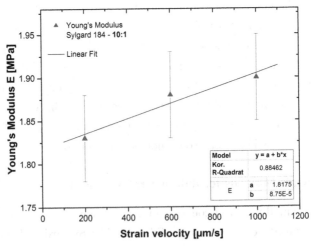

Figure 4.10 – Dependency of the Young's modulus on the strain-velocity of Sylgard 184 with a mixing-ratio of 10:1.

However, a certain deviation of the Young's modulus could be observed as shown in Figure 4.10, indicating an increasing material stiffness with higher strain-rates. In particular the graph yields an increase of ≈4% for the mentioned variation of the strain-velocity, which is in good agreement to the findings of Schneider et al. [90].

4.2 Determination of the Resistance of Stretchable Electrodes

The following section presents resistance over strain measurements of thin metallic layers on PDMS with different electrode widths. The obtained results are compared with measurements carried out on commercial equipment. In particular chromium-gold bilayers (5/50nm) with electrode widths of 1mm and 0.11mm on 1mm thick PDMS substrates are characterized and compared to the results presented by Graz et al. [79]. The samples were fabricated as described in this paper.

4.2.1 Sample Preparation

Similar to the preparation of the samples used for stress-strain measurements, PDMS films with 1mm uniform thickness were fabricated. On the substrate dog-bone shaped bilayers consisting of a 5nm chromium interface layer and a 50nm gold layer were deposited. The Cr layer ensures proper adhesion of the gold to the PDMS. The 4x4mm pads on both sides of the electrodes are used for contacting the electrode. Figure 4.11(a) shows the geometry of the electrodes. The electrodes were contacted by using copper wire, conductive silver paste and 4x4mm PDMS pads. To ensure a permanent contacting also two 4x4mm magnets were applied as shown in Figure 4.11(b).

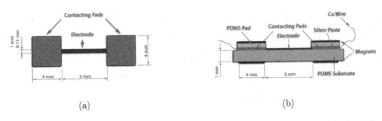

(a)   (b)

Figure 4.11 – a) Geometry of the dog-bone shaped Cr/Au (5/50nm) thin-film electrodes. b) Electrode contacting with copper wire, silver paste, PDMS pads and fixation magnets.

The metallization was done by thermal evaporation through a shadow mask. This allows for a reproducible and homogeneous deposition of the material and good control of the layer thickness. Both layers were deposited with a rate between 0.03 and $0.1\frac{nm}{s}$. This ensures the formation of small grains and microcracks, which account for the persistive conduction during stretch as discussed in [79].

4.2.2 Substrate Geometry

The stretchable conductors need a slightly different treatment when strained than pure PDMS samples. Since the electrode has a certain contribution to the "Young's modulus" of such an elastomer-metal combination, only the part where the actual electrode is located has to be taken into consideration for strain measurements. Otherwise mainly the electrode surroundings tend to be stretched rather than the electrode itself. Since the setup is only capable of measuring the grip-grip displacement, it is worthwhile that the grips are located closely to the "interesting" part of the sample, the electrode. Therefore, the strain of such a sample can be measured by simply taking the grip-grip displacement with the initial grip-displacement as reference length without further correction.

The uniformity of straining a cuboid sample was tested similar as already done with the ISO527-3 conform clamping, but with the difference that the sample was clamped in the direct adjacency of the cuboid shaped bar, the formerly called "uniform region". This $L_0 = 10mm$ long part was again taken as reference and the strain $\varepsilon_{uniform}$ measured manually. Figures 4.12(a) and (b) show the measured strain according to the grip-grip displacement ε_{total} with the initial displacement L_{init} taken as reference.

$$\varepsilon_{total} = \frac{\Delta L}{L_{init}} \qquad \varepsilon_{uniform} = \frac{\Delta L_0}{l_0} \tag{4.3}$$

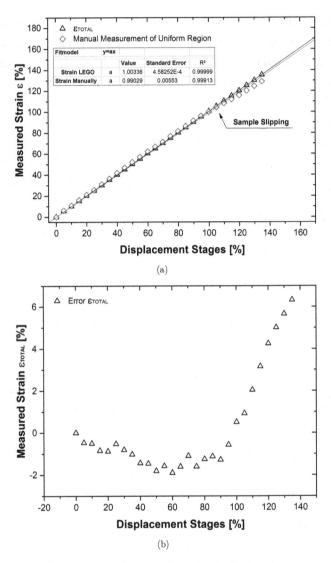

(a)

(b)

Figure 4.12 – a) Sample strain determined via the grip displacement ε_{total} compared to the manual measurement in the center region of the cuboid sample $\varepsilon_{uniform}$ b) Deviation of the measured strain to the strain in the center region.

From the results shown in the latter figures it can be deduced that also cuboid samples can also be used. Nevertheless, the uniform stress distribution is not guaranteed, since the squeezed grip adjacency still creates a highly inhomogeneous region. However for a qualitative assessment and as basis for comparison of different conductors this strain-measurement method is sufficient.

For the measurements a rectangular shaped sample with 35x10mm as shown in Figure 4.13(a) was used. The samples were clamped closely to the contacting pads with an initial grip-grip displacement of 19mm. Figure 4.13(b) shows the sample applied to the setup.

(a)   (b)

Figure 4.13 – a) Flexible electrode sample 5/50nm Cr/Au bilayer with an electrode width of 110μm on 35x10x1mm 10:1 PDMS. b) Sample clamped in the measurement device and contacted with copper wires, silver paste and 4x4mm PDMS pads. Proper contacting was assured by four magnets.

4.3 Measurement Procedure

The samples were cycled with $50\frac{\mu m}{s}$ to a maximum strain of $\varepsilon = 20\%$ in steps of 1% and with a rest-time of 60s at every step. The resistance was acquired permanently. Prior to each measurement the initial resistance was measured with a digital multimeter U1237A (Agilent) and compred to the readings from the measurement setup.

In Figures 4.14 the first cycle for the 0.11mm electrode is shown. The initial resistance R_0 of 699Ω was also confirmed with the digital multimeter. The ascent of the resistance over strain up to 20% is almost linear with a maximum resistance R_{MAX} of 1583Ω. The relaxation sweep (20% $\rightarrow$ 0%) clearly shows, that the conducting characteristics improve after the first stretching cycle, since the curve runs significantly under the stretching curve and results in a final resistance $R_F = 599\Omega$.

The second and also the following cycles do not show this behaviour, but almost return to the initial resistance. However, the resistance tends to increase with every following cycle. Figures 4.15(a) and 4.15(b) show the second cycle of the same electrode which is also the basis for comparison with [79], since here also the second stretching cycle is shown as representative measurement. The resistance for 0% remains almost unchanged since R_0 and R_F are 599Ω and 588Ω respectively. The maximum resistance R_{MAX} at 20% strain was 1370Ω.

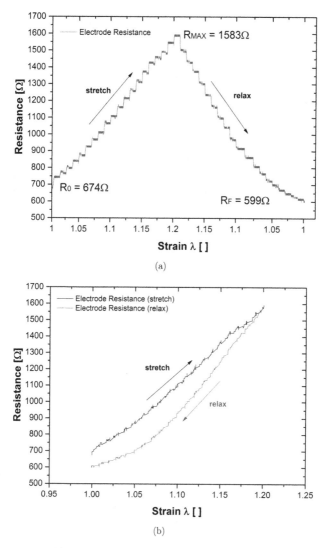

(a)

(b)

Figure 4.14 – a) First resistance over strain cycle for the 110μm electrode, with a maximum strain of 20%. The stretching was done in 1% steps with a waiting time of 60s at every step and a strain-velocity of $50\frac{\mu\text{m}}{\text{s}}$. b) Same data shown in the hysteresis plot.

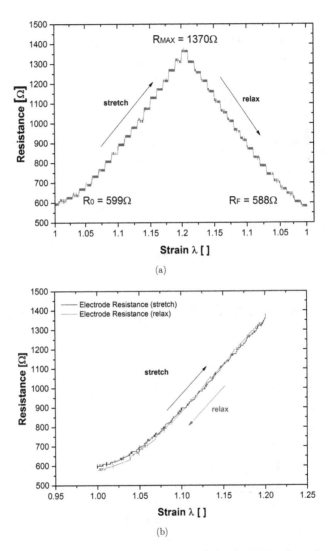

Figure 4.15 – a) Second resistance over strain cycle for the 110μm electrode, with a maximum strain of 20%. The stretching was done in 1% steps with a waiting time of 60s at every step and a strain-velocity of $50\frac{\mu m}{s}$. b) Same data shown in the hysteresis plot.

Secondly also the 1mm electrode was measured identically to the latter. The electrode improvement during the first cycle could also be confirmed for that electrode. The resistance in the unstrained state drops from 50Ω

to 42Ω. For comparison with [79], Figures 4.16(a) and (b) show the second
stretching cycle.

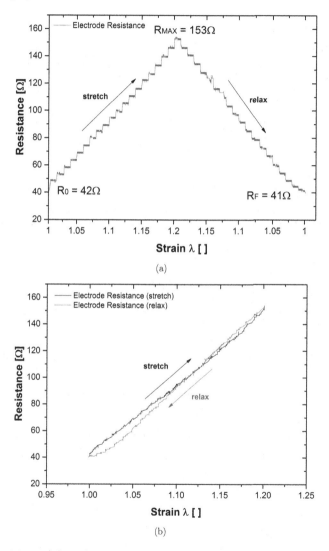

(a)

(b)

Figure 4.16 – a) Second resistance over strain cycle for the 1mm electrode, with a
maximum strain of 20%. The stretching was done in 1% steps with a waiting time of
60s at every step and a strain-velocity of $50\frac{\mu m}{s}$. b) Same data shown in the hysteresis
plot.

The results are in good agreement to the findings presented by Graz et al. in [79]. Tables 4.5 and 4.6 show a comparison of the obtained resistance progression.

Table 4.5 – Comparison of the findings of Graz et al. with the determined results shown in the latter figures for the 1mm electrode. The second stretching cycle was used as basis for comparison.

	LEGO setup	Graz et al.
R_0 $[\Omega]$	42	50
R_{MAX} $[\Omega]$	153	176
R_F $[\Omega]$	41	48

Table 4.6 – Comparison of the findings of Graz et al. with the determined results shown in the latter figures for the 110 μm electrode. The second stretching cycle was used as basis for comparison.

	LEGO setup	Graz et al.
R_0 $[\Omega]$	599	860
R_{MAX} $[\Omega]$	1370	1536
R_F $[\Omega]$	588	648

The resistances and overall behaviour of the strain progression are almost identical, the differences can be explained by manufacturing irregularities. The trend could be reproduced in good agreement. Furthermore, the control readings from the digital multimeter were identical to the measurement of the LEGO setup, confirming the correct measurement of the resistance. This proofs the capability of the device for carrying out trustable resistance-strain measurements.

5 Conclusion & Outlook

"We are at the very beginning of time for the human race. It is not unreasonable that we grapple with problems. But there are tens of thousands of years in the future. Our responsibility is to do what we can, learn what we can, improve the solutions, and pass them on."

— Richard P. Feynman

The LEGO®-based measurement device presented in this work fulfils the demanded requirements, in particular offering an uncommon, simple and cost effective approach to scientific measurement by using LEGO® Mindstorms and self-built "smart-sensors". The setup is capable of performing uniaxial extension up to 30mm with velocities up to 1300 $\frac{\mu m}{s}$. Strain is measured with an accuracy of 100μm by determining the grip-grip displacement with a digital sliding calliper. Synchronous determination of tensile force or resistance to the strain enables for stress-strain and resistance-strain measurements. For stress-strain studies a maximum tensile force of 35N with an accuracy of $\leq$0.03N is possible. Resistance-strain measurements can be conducted in a range from 0-1MΩ, while offering different measurement ranges to ensure a high accuracy throughout the whole range. Furthermore, an intuitive LabView user-interface enables for real-time tracking of the measurement along with permanent control of the device.

To evaluate the accuracy and functionality of the LEGO®-tensometer, PDMS with varying mixing-ratio and thin-film gold electrodes on PDMS were characterized. The data was compared to the data obtained with commercial equipment by Schneider et al. [90] and Graz et al. [79], with remarkable agreement. With these results we have successfully shown that LEGO® is not only a kid's toy but can be a powerful tool in scientific research. Being available almost world-wide, it is an ideal platform not only for cost-effective rapid-prototyping and assisting tasks, but also for conduction of sophisticated and trustable measurements. Due to it's compatibility and simplicity it allows people with only basic engineering knowledge to reproduce even complicated designs, which is interesting for educational purposes eg. project-based classes. Furthermore, it shows that equipment for scientific research does not always have to be expensive. Financial means are often limited and with the upcoming ingress of less-developed countries into scientific research, cost-effective solutions become more and more important. Surely, in many cases costly equipment is needed for scientific research, but it is often not necessary. Science should be more about the use of creativity and improvisation for going forward and not about which tools you use to get there.

The work was recently presented at the EuroEAP 2013 conference in Dübendorf (Zürich), Switzerland and awarded with the *Best Poster Award*.

EuroEAP 2013
International conference on
Electromechanically Active Polymer (EAP)
transducers & artificial muscles
Dübendorf (Zürich), 25-26 June 2013

Organized and supported by
"European Scientific Network for Artificial
Muscles – ESNAM' (www.esnam.eu)
COST Action MP1003

Poster ID:

1.2.16

Contact e-mail:
Richard.Moser84@gmail.com

Plastic tests plastics: A toy brick based tensometer for the characterization of dielectric elastomers

Richard Moser (1), Ingrid M. Graz (1), Christian M. Siket (1), Gerald Kettlgruber (1),
Petr Bartu (1), Michael Drack (1), Siegfried Bauer (1), Umut D. Cakmak (2), Zoltan Major (2)

(1) Johannes Kepler University, Soft Matter Physics, Linz, Austria; (2) Johannes Kepler University, Institute of Polymer Product Engineering, Linz, Austria;

Abstract

Knowledge of stress-strain curves of dielectric elastomers is important for the design of dielectric elastomer actuators and generators, usually displayed in work-conjugate plots. Measurements of stress strain curves are typically carried out with expensive commercial tensometers. Here we show how to use and upgrade toy bricks based on the Lego® Mindstorms system for the construction of a lightweight, low-cost and easy to reproduce tensometer. The current design allows for stress-strain studies along with resistance over strain measurements, therefore being an ideal tool for mechanical characterization of common elastomers and performance studies of stretchable electrodes. We apply our system to mechanically characterize PDMS elastomers and compare results with measurements obtained on commercial equipment, with remarkable agreement. Additionally we show hot to use the set-up for the electromechanical characterization of stretchable electrodes, based on thin metal layers on dielectric elastomers.

LEGO®, just a kid's toy?

- Available world wide
- Cost-effective rapid prototyping
- Interesting for scientific and educational purposes
- Powerful tool in scientific research

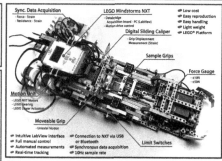

Sync. Data Acquisition — Force - Strain, Resistance - Strain
- Databridge
- Acquisition board – PC (LabView)
- Motion drive control

LEGO Mindstorms NXT
- Low cost
- Easy reproduction
- Easy handling
- Light weight
- LEGO® Platform

Digital Sliding Caliper — Grip Displacement Measurement (Strain)

Sample Grips

Force Gauge — ±10N, ±50N

Motion Unit
- LEGO NXT Motors
- LEGO Gearing
- LEGO Cluster Actuation

Moveable Grip — Uniaxial Motion

- Intuitive LabView Interface
- Full manual control
- Automated measurements
- Real-time tracking

- Connection to NXT via USB or Bluetooth
- Synchronous data acquisition
- 10Hz sample rate

Limit Switches

Mechanical Characterization
of elastomers

- Stress-Strain Studies
- Determination of Young's Modulus (E)
- ISO527-5B "Dog-Bone" shaped samples
- Polydimethylsiloxane (PDMS)
- Sylgard 184 (Dow Corning)

Electrical Characterization
of stretchable conductors

- Resistance-Strain Studies
- Stretchable electrode performance
- Cuboid shaped PDMS 10:1 substrate
- 5nm Cr/ 50nm Au Bi-layer (PVD)
- Electrode 10x1 mm

Tensile Force	Strain	Resistance
- Vernier® Force Gauge	- Digital sliding caliper	- 0-1MΩ
- Dual Range	- max. 30mm linear motion	- 5 measurement ranges
- 10N; Accuracy: 0.03N	- Grip-Grip displacement	- Accuracy: 0.1%
- 50N; Accuracy: 0.15N	- Accuracy: 100μm	

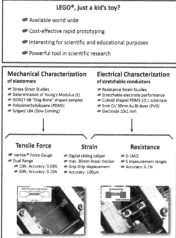

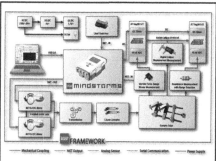

FRAMEWORK

— Mechanical Coupling   — NXT Output   — Analog Sensor   — Serial Communication   — Power Supply

Acknowledgments

Participation to this conference was partially supported by COST (European Cooperation in Science and Technology) in the framework of ESNAM (European Scientific Network for Artificial Muscles) - COST Action MP1003.

References

1. F. Schneider, T. Fellner, J. Wilde, and U. Wallrabe, *Mechanical properties of silicones for MEMS*, Journal of Micromechanics and Microengineering, vol. 18, no. 6, p. 065008, 2008.
2. I. M. Graz, D. P. Cotton, and S. P. Lacour, *Extended cyclic uniaxial loading of stretchable gold thin-films on elastomeric substrates*, Applied Physics Letters, vol. 94, no. 7, pp. 071 902071 902, 2009

6 Appendix

This section contains the full schematics, board design and microcontroller firmware for the intelligent LEGO compatible sensors. The schematics and the circuit board were drawn with *Eagle 5.0* and the board was fabricated with in-house photo lithography and etching. The firmware was written in ANSI C using the compiler *MicroC Rev. 5.0*. Programming of the ATTiny261 controllers was done with a *mikroElektronika - EasyAVR5* development board. Additionally the building instruction plan for the LEGO framework is included and also available as interactive HTML interface. The model was built in LEGO Digital Designer 4.6.3. Furthermore, the CAD design of the sample sledge done in *SolidWorks 2012* is attached.

6.1 Intelligent Sensor Schematics

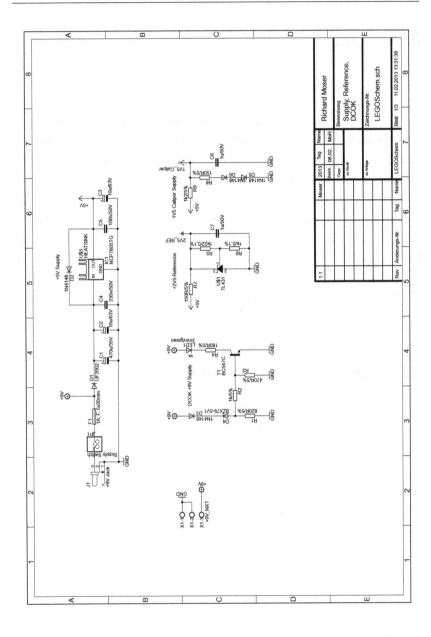

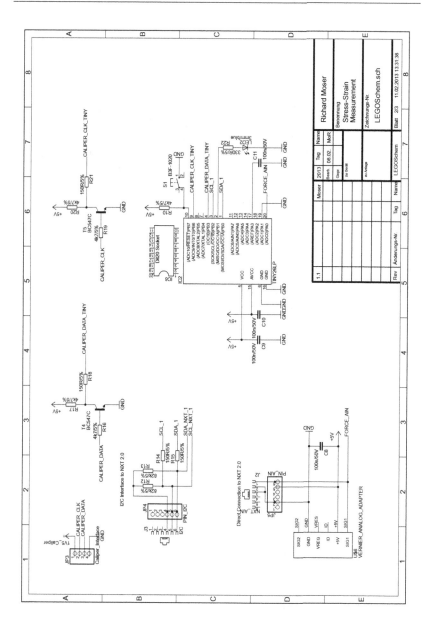

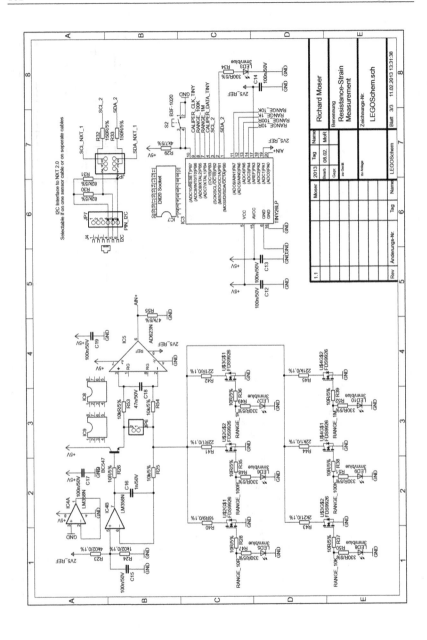

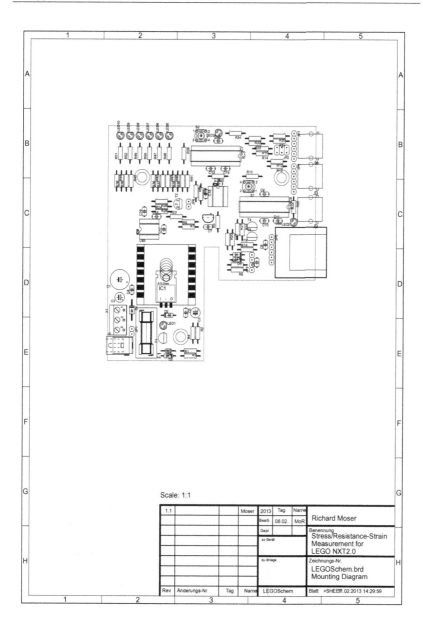

Scale: 1:1

1.1				Moser	2013	Tag	Name	Richard Moser
				Bearb	08.02.	MoR		
				Gepr.			Benennung	Stress/Resistance-Strain Measurement for LEGO NXT2.0
				zu Gerät				
				zu Anlage			Zeichnungs-Nr.	LEGOSchem.brd Mounting Diagram
Rev	Änderungs-Nr.	Tag	Name	LEGOSchem			Blatt >SHEET 11.02.2013 14:29:59

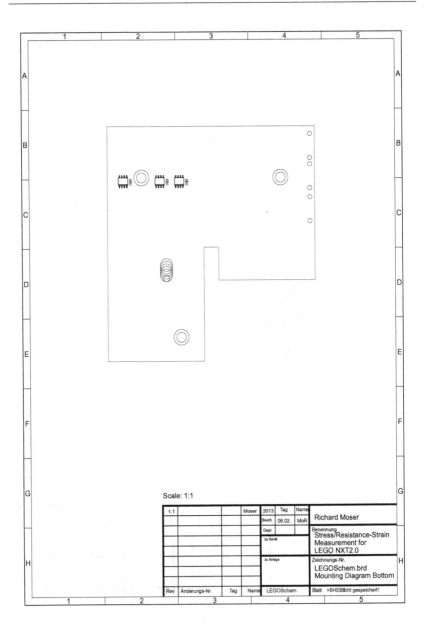

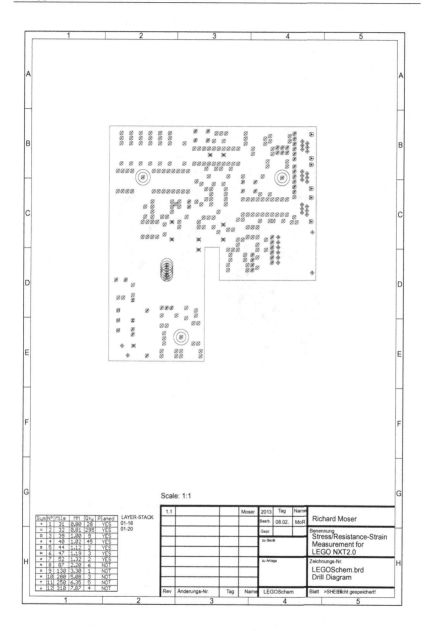

Scale: 1:1

Sym	N°	Mils	MM	Qty	Plated
+	1	31	0.80	28	YES
×	2	32	0.81	295	YES
□	3	39	1.00	9	YES
◇	4	40	1.02	45	YES
⊠	5	44	1.12	2	YES
⋈	6	47	1.19	3	YES
◆	7	52	1.32	2	YES
◆	8	87	2.20	6	NOT
×	9	130	3.30	1	NOT
⋈	10	200	5.08	3	NOT
▼	11	250	6.35	5	NOT
▲	12	310	7.87	4	NOT

LAYER-STACK
01-16
01-20

1.1			Moser	2013	Tag	Name
			Bearb	08.02.	MoR	Richard Moser
			Gepr.			Benennung
			zu Gerät			Stress/Resistance-Strain Measurement for LEGO NXT2.0
			zu Anlage			Zeichnungs-Nr.
						LEGOSchem.brd
Rev	Änderungs-Nr.	Tag	Name	LEGOSchem		Blatt >SHEET nicht gespeichert!

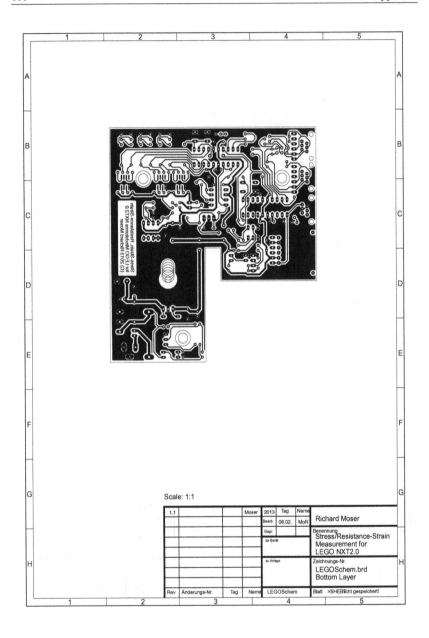

Scale: 1:1

1.1				Moser	2013	Tag	Name	
				Bearb.	08.02.	MoR		Richard Moser
				Gepr.				Benennung
				zu Gerät				Stress/Resistance-Strain Measurement for LEGO NXT2.0
				zu Anlage				Zeichnungs-Nr. LEGOSchem.brd Bottom Layer
Rev	Änderungs-Nr.	Tag	Name	LEGOSchem			Blatt	>SHEET<icht gespeichert!

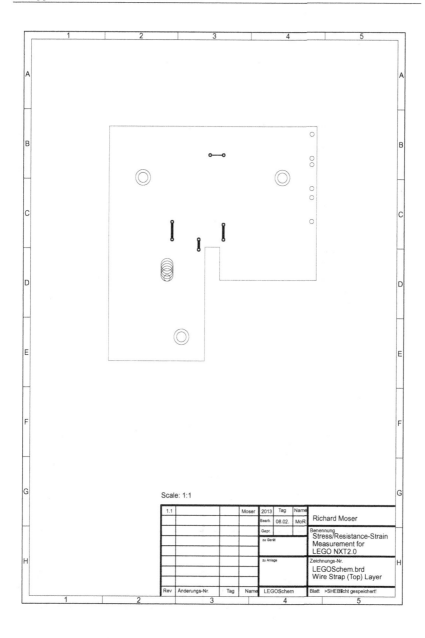

Scale: 1:1

1.1			Moser	2013	Tag	Name	Richard Moser
				Bearb.	08.02.	MoR	
				Gepr.			Benennung
				zu Gerät			Stress/Resistance-Strain Measurement for LEGO NXT2.0
				zu Anlage			Zeichnungs-Nr.
							LEGOSchem.brd Wire Strap (Top) Layer
Rev	Änderungs-Nr.	Tag	Name	LEGOSchem		Blatt	>SHEET<nicht gespeichert!

1	2	3		4	5
Item #	Qty.	Description		Part ID	Distributor
1	1	RESISTOR	820R / 5% / 0.25W	R1	RS-Components
2	1	RESISTOR	1k5 / 5% / 0.25W	R2	RS-Components
3	1	RESISTOR	470R / 5% / 0.25W	R3	RS-Components
4	1	RESISTOR	180R / 5% / 0.25W	R4	RS-Components
5	2	RESISTOR	1k02 / 0.1% / 0.25W	R5, R24	RS-Components
6	1	RESISTOR	1k0 / 0.1% / 0.25W	R6	RS-Components
7	8	RESISTOR	150R / 5% / 0.25W	R7, R8, R14, R15, R18, R21, R32, R33	RS-Components
8	1	RESISTOR	1k2 / 5% / 0.25W	R9	RS-Components
9	4	RESISTOR	82k / 5% / 0.25W	R12, R13, R30, R31	RS-Components
10	5	RESISTOR	4k7 / 5% / 0.25W	R10, R16, R17, R19, R20, R29	RS-Components
11	8	RESISTOR	330R / 5% / 0.25W	R22, R47, R48, R49, R50, R51, R52, R34	RS-Components
12	1	RESISTOR	4k02 / 0.1% / 0.25W	R23	RS-Components
13	1	RESISTOR	10R / 5% / 0.25W	R26, R28, R35, R36, R37, R38, R39	RS-Components
14	3	RESISTOR	10k / 5% / 0.25W	R25, R53, R54	RS-Components
15	1	RESISTOR	47k / 5% / 0.25W	R55	RS-Components
16	1	RESISTOR	10k / 5% / 0.25W	R25, R53, R54	RS-Components
17	3	RESISTOR	16R9 / 0.1% / 0.25W	R40	RS-Components
18	1	RESISTOR	22R1 / 0.1% / 0.25W	R41	RS-Components
19	1	RESISTOR	221R / 0.1% / 0.25W	R42	RS-Components
20	1	RESISTOR	2k21 / 0.1% / 0.25W	R43	RS-Components
21	1	RESISTOR	22k1 / 0.1% / 0.25W	R44	RS-Components
22	1	RESISTOR	221k / 0.1% / 0.25W	R45	RS-Components
23	1	ELEC. CAP	470µ / 35V	C1	RS-Components
24	2	ELEC. CAP	10µ / 63V	C2, C3	RS-Components
25	1	CER. CAP	330n / 50V	C4	RS-Components
26	11	CER. CAP	100n / 50V	C5, C8, C9, C10, C11, C12, C13, C14, C15, C17, C19	RS-Components
27	1	CER. CAP	1µ / 50V	C6, C7	RS-Components
28	1	CER. CAP	1n / 50V	C16	RS-Components
29	1	CER. CAP	47n / 50V	C18	RS-Components
30	1	DIODE	UF2002	D1	RS-Components
31	4	DIODE	1N4148	D2, D3, D5, D6	RS-Components
32	1	Z-DIODE	BZX79-5V1	D4	RS-Components
33	1	LED	3mm/green	LED1	RS-Components
34	8	LED	3mm/blue	LED2, LED3, LED5, LED6, LED7, LED8, LED9, LED10	RS-Components
35	1	TRANSISTOR	BC547C	T1, T4, T5, T2	RS-Components
36	3	MOSFET	FDS9926 / dual N	U2, U3, U4	RS-Components
37	1	REFERENCE	TLV431	U1	RS-Components
38	1	V-REG	LM7805	IC1	RS-Components
39	2	µCONTROL	ATTiny261	IC2, IC3	RS-Components
40	1	OPAMP	LM358	IC4	RS-Components
41	1	INSTR. AMP	AD623	IC5	RS-Components

				Name: Richard Moser, BSc.	
			2013	Date	Name	Description: LEGO Mindstorms NXT 2.0	
			Bearb:	07.02.	Moser R.	Measurement Device:
			Gepr:			Stress-Strain / Resistance-Strain
			Normg:			

JKU Linz	Department: SOMAP	Document ID.: PartsList.docx	Page: 1 1 Pg.
		Replaced by: -	Substitute for: -

1	2	3	4	5
Item #	Qty.	Description	Part ID	Distributor
39	1	DC-JACK	J1	RS-Components
40	1	DC-JACK 3-Terminal	X1	RS-Components
41	1	ADAPTER Vernier analog sensor-adapter	U6	www.vernier.com
42	3	NXT2.0 JACK	J2, J3, J4	RS-Components
43	2	PIN CONNECT 1x2 / 2.54mm	JP1, JP6	RS-Components
44	1	PIN CONNECT 1x4 / 2.54mm	JP3	RS-Components
45	3	PIN CONNECT 1x6 / 2.54mm	JP4, JP5, JP7	RS-Components
46	1	PIN CONNECT 2x3 / 2.54mm	JP2	RS-Components
47	3	JUMPER 2-Terminal / 2.54mm	-	RS-Components
48	2	SOCKET DIL8	IC8, IC9	RS-Components
49	3	SOCKET DIL20	IC6, IC7	RS-Components
50	1	FUSEHOLDER 5x20mm / RM 22,5mm	F1	RS-Components
51	1	FUSE 5x20mm / 1A / F	F1	RS-Components
52	2	MICROBUTTON SPST 5mm	S1, S2	RS-Components
53	1	HEATSINK TO220 13K/W	U5	RS-Components
54	1	CIRC. BOARD Bungard / 160x200	-	Distrelec
55	1	DC SUPPLY 12V / 1,5A / stab.	-	RS-Components
56	1	CABLE 4-Terminal Digital Caliper Interface	-	RS-Components
57	1	FORCE GAUGE Vernier Force- Gauge ±10N / ±50N	-	www.vernier.com
58	1	DIG. CALIPER WISENT dig. Caliper	-	Bauhaus
59	1	LEGO LEGO Mindstorms NXT2.0 8547	-	LEGO Inc.
60	1	LEGO LEGO Excavator 8294	-	LEGO Inc.
61	1	CLAMP DEVICE	-	-
62	1	TORQUE GAUGE WERA 7440 ESD 0,3-1,2Nm	-	RS-Components
63	1	SOFTWARE NI LabView 2011	-	www.NI.com

Klasse:		Kat. Nr.:	Name:
			Richard Moser, BSc.

2013	Date	Name	Description:
Bearb:	07.02.	Moser R.	LEGO Mindstorms NXT 2.0
Gepr:			Measurement Device:
Normg:			Stress-Strain / Resistance-Strain

JKU Linz	Department: SOMAP	Document ID.: PartsList.docx	Page: 1 1 Pg.
		Replaced by: -	Substitute for: -

6.2 Microcontroller Firmware

The following section gives a brief overview of the ATTiny firmware for the synchronous acquisition of stress, strain and resistance and the I^{2}C communication to the Mindstorms NXT.

Since the two firmwares (stress-strain, resistance-strain) are very similar, its not useful to attach both, so only the resistance-strain firmware is attached. It contains every part of the stress-strain firmware, just with the additional functionality of supporting different measurement ranges. The parts, only found in the resistance-strain sensor are highlighted in the flow-charts and marked as "res-strain only" in the source code.

The functionality itself is completely interrupt controlled. The clock from the digital calliper interface triggers the external interrupt initializing the pick-up of a new displacement reading. Parallel to this, four ADC samples are taken and averaged. This ensures a synchronous displacement-ADC value data-point. If the NXT, connected via I^{2}C (or TWI for ATMEL controllers) requests data, first the configuration data (desired resistance-measurement-range) from the NXT is taken and the requested measurement range is set by reconfiguring the precision current source. Afterwards the current data-point is copied into the transmit-buffers and sent to the NXT.

6.2.1 Main Routine

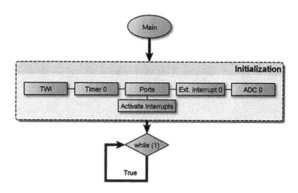

Figure 6.1 – Flowchart of the firmware main routine.

The rain routine represents the entering point of the firmware. Prior to the endless main-loop the controller is initialized and the used functions (dig. ports, timer interrupt, I²C ID, TWI-interface, external interrupt 0 and ADC) are configured. In the main-loop the controller waits for a data-request from the NXT.

```
1  /* ********************************************************************
2  *
3  * Somap Corporation
4  *
5  * File          : USI_TWI_Slave.h
6  * Compiler      : MicroC PRO
7  * Revision      : Revision: 5.00
8  * Date          : Date: Monday, January 31, 2013
9  * Updated by    : Author: Richard Moser
10 *
11 * Supported devices : ATTiny261
12 *
13 * Description   : Main file. Application picks up the Caliper Protcol,
14 *                 generates the displacement value and transfers it (2
       Byte)
15 *                 via TWI/I2C communication to the LEGO NXT2.0 brick.
16 *                 Furthermore the ADC is sampled synchronously to the
       caliper
17 *                 acquisition, averaged and also sent via I2C.
18 *                 The ATTiny acts as TWI Slave in this configuration
19 *                 where the TWI communication routines are based on the
20 *                 ATMEL AVR312 Application Note.
```

```
21 *
22 ********************************************************************/
23
24 #include "USI_TWI_Slave.h"
25 #include "Lego_Caliper_Slave.h"
26
27 /* The main function.
28  * The program entry point. Initates the controller and enters eternal
        loop, waiting for data.
29  */
30
31 void main( void ){
32
33   unsigned char TWI_slaveAddress;
34   unsigned int act_caliper_data = 0;
35   char *cpoint;
36
37   // *** LED feedback port
38   DDRB = 0xFF; // Set to output
39   DDRA = 0xF4; // PA0 und PA1 for ADC (diff.), the others - dig.
        output
40                   // PA3 AREF Pin
41
42   // *** Configure Digital Caliper Ports
43   DDB6_bit = 0;          // set PB6 (INT0) as input (Caliper_Clock)
44   PORTB6_bit = 1;        // switch on internal Pullup Resistor
45   DDB3_bit = 0;          // set PA6 as input (Caliper_Data)
46   PORTB3_bit = 1;        // switch on internal Pullup Resistor
47   DDB1_bit = 1;          // control led
48   PORTB1_bit = 1;        // reset inicator Bit
49
50   // *** Configure Timer Interrupt
51   TOIE0_bit = 1;         // Timer0 overflow interrupt enable
52   TCCR0A = 0;            // No special timer functions
53   TCCR0B = 3;            // Start timer with 64 prescaler
54   TCNT0L = 0;            // Initialize Timervalue
55   TCCR1B = 7;            // Start timer with 64 prescaler
56
57   // *** Configure I2C Slave Adress, Initialize I2C Bus
58   TWI_slaveAddress = 0x20; // Own TWI slave address
59   USI_TWI_Slave_Initialise( TWI_slaveAddress );
60
61   // *** Configure External Interrupt
62   INT0_bit = 1;          // Enable external Interrupt 0
63   ISC00_bit = 0;         // falling Edge of INT0 generates Interrupt
64   ISC01_bit = 1;
65
66   // *** Configure ADC
67   // Set ADC reference to external reference, PA0 and PA1 as
        differential, gain 1x/8x
```

```
68   ADMUX |= (0 << REFS1)|(1 << REFS0)|(1 << MUX0)|(0 << MUX1)|(0 <<
     MUX2)|(0 << MUX3)|(0 << MUX4);
69
70   // Set ADC prescaler to 128;62.5kHz sample rate @ 8MHz, Enable ADC,
     No automatic Trigger, ADC Interrupt Enable
71   ADCSRA |= (0 << ADATE)|(1 << ADPS2)|(1 << ADPS1)|(0 << ADPS0)|(1 <<
     ADIE);
72
73   // Select 2.56V (1) or 1.1V (0) internal Reference Voltage, select 1
     x Gain
74   ADCSRB |= (0 << REFS2)|(1 << MUX5)|(0 << GSEL);
75   ADCSRA |= (1 << ADEN);
76
77   // *** Configure SleepMode
78   MCUCR |= (1 << SM0)|(0 << SM1);
79
80   SREG_I_bit = 1;          // Global interrupt enable
81   asm{SEI};                // enable interrupts
82
83   // *** This loop runs forever.
84   while(1)
85   {
86     // check if new data arrived via I2C
87     get_TWI_ReceiveBuffer();
88   }
89 }
```

6.2.2 Data Acquisition Routine

This routine is responsible for picking up the data from the connected sensors, the digital calliper and either resistance or force. However, the acquisition of resistance and force is equal since only the ADC reading is recorded. This raw value is transferred to the NXT and bridged to the LabView program, where the final quantity is determined from this raw-value.

To get the data mainly three interrupt service routines are necessary. The external interrupt *External0_ISR*, the timer interrupt *Timer0Overflow_ISR* and the ADC interrupt *ADC0_ISR*. The external interrupt is always triggered if a falling edge on the correspondent controller pin is detected. On every trigger the corresponding calliper data bit (always being valid at the falling clock edge) is shifted with into the variable *shift-data*. Additionally on the first four interrupts the ADC is sampled, whereas on completion the ADC interrupt is called adding up the acquired values in the variable *ADC_act_data_int*. Furthermore, with every detected calliper clock timer0 is reset to prevent calling the correspondent interrupt routine. If finally all 24 calliper bits were read the timer over-flows and the timer0 interrupt routine is called. Since new calliper data is available, the data is reduced to two bytes, the sign-bit is handled and the current position value finally stored in *act_data2Byte*. Next the four read ADC values are averaged and the resulting 10-bit ADC value is copied into the two bytes *ADC_data_LByte* and *ADC_data_HByte*. Lastly the resistance measurement range is reconfigured (if necessary) by activating the corresponding reference resistor in the precision current source and re-configuring the ADC gain. Furthermore, it is important that during this acquisition procedure the TWI-interrupts are disabled, meaning that all requests from the NXT are discarded during the data acquisition. This is necessary, since it was found that the interrupts highly disturb each other which would result in errors.

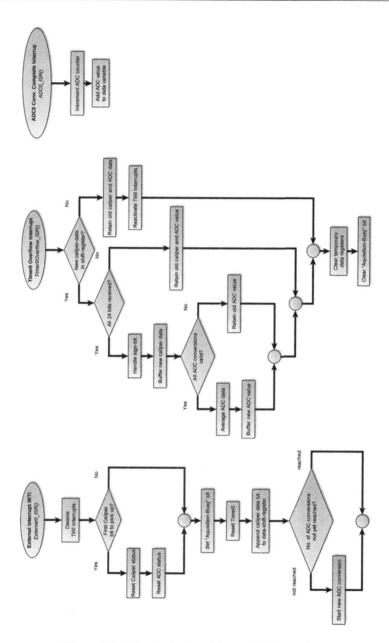

Figure 6.2 – Flowchart of the data acquisition routine.

6.2.2.1 Lego_Caliper_Slave.h

```
1 /* ********************************************************************
2 *
3 * Somap Corporation
4 *
5 * File          : Lego_Caliper_Slave.h
6 * Compiler      : MicroC PRO
7 * Revision      : Revision: 5.00
8 * Date          : Date: Monday, November 05, 2012
9 * Updated by    : Author: Richard Moser
10 *
11 * Supported devices : ATTiny261
12 *
13 *
14 * Description   : Header file for readout of Wisent digital Caliper
15 *
16 ********************************************************************/
17 //! Prototypes
18 unsigned int get_act_caliper_value ( void );
19 unsigned char get_act_ADC_HByte ( void );
20 unsigned char get_act_ADC_LByte ( void );
21 void get_TWI_ReceiveBuffer ( void );
```

6.2.2.2 Lego_Caliper_Slave.c

```
1 /********************************************************************
2 *
3 * Somap Corporation
4 *
5 * File          : Lego_Caliper_Slave.c
6 * Compiler      : MicroC PRO
7 * Revision      : Revision: 5.00
8 * Date          : Date: Monday, November 05, 2012
9 * Updated by    : Author: Richard Moser
10 *
11 * Supported devices : ATTiny261
12 *
13 *
14 * Description   : Source file for readout of Wisent Digital Caliper
15                   and ADC value synchronous to Caliper signal
16                   Includes ISR Timer, External0 and ADC Interrupt.
17 *
18 ********************************************************************/
19
20 #include "Lego_Caliper_Slave.h"
21 #include "USI_TWI_Slave.h"
22
23 #define ADC_AVERAGING    4    // averaged ADC values
```

```
24 #define AVERAGING_SHIFT  2    // bits to shift to average all ADC
     values
25
26 char counter = 0;
27 unsigned long shift_data = 0;
28 unsigned long act_data = 0;
29 unsigned int act_data2Byte = 0;
30 unsigned int act_data2Byte_temp = 0;
31 unsigned char ADC_data_LByte = 0;
32 unsigned char ADC_data_HByte = 0;
33 unsigned int ADC_HByte_int = 0;
34 unsigned int ADC_LByte_int = 0;
35 unsigned int ADC_act_data_int = 0;
36 unsigned int ADC_data_int_average = 0;
37 unsigned char ADC_conversion_count = 0;
38 char caliper_count = 0;
39 unsigned char Resistance_Range = 3;
40 unsigned char ManRange = 10;
41
42 // Resistance Range:
43 // 0: 10R
44 // 1: 100R
45 // 2: 1000R
46 // 3: 10.000R
47 // 4: 100.000R
48 // 5: 1.000.000R
49 // 10: Automatic Selection
50
51 unsigned int get_act_caliper_value ( void ) {
52   return act_data2Byte;
53 }
54
55 unsigned char get_act_ADC_HByte ( void ) {
56   return ADC_data_HByte;
57 }
58
59 unsigned char get_act_ADC_LByte ( void ) {
60   return ADC_data_LByte;
61 }
62
63 void get_TWI_ReceiveBuffer ( void ){
64   if( USI_TWI_Data_In_Receive_Buffer() ){
65     ManRange = USI_TWI_Receive_Byte();
66   }
67 }
68
69 void Timer0Overflow_ISR() org IVT_ADDR_TIMER0_OVF {
70   /* ***
71     On overflow it's checked if the caliper acquisition routine got
        data.
```

72   If so, the data is copied into the I2C-send registers, from where
      they are sent to the NXT.
73   Furthermore, it's checked if the previous resistance measurement
      was done in the optimal range (if auto is selected), and the
      range is changed for the next measurement if necessary. On
      manual range selection the range is set according to the
      request.
74   TWI Interrupts are reactivated.
75   *** */
76
77   /* *** Pause detected -- set new value if 24 bits detected *** */
78   if ((PINB1_bit == 0)){
79   if(caliper_count == 24){   // 24 bits received from
        caliper
80   shift_data = shift_data >> (32-caliper_count); // shift to
          beginning of 32bit variable
81   act_data = shift_data;   // copy new value
82   act_data2Byte_temp = (unsigned int) (0xFFFF & act_data);
83   // ignore sign-bit (1st bit of last 4Bit packet)
84
85   // add sign bit to 2Byte data, located in byte 3 of act_data
86   if(act_data & 0x100000){
87   act_data2Byte_temp = act_data2Byte_temp | 0x8000;
88   }
89   else { }
90
91   /* *** save results *** */
92   // caliper
93   act_data2Byte = act_data2Byte_temp;
94
95   // ADC
96   if(ADC_conversion_count >= (ADC_AVERAGING)){   // check if
          every conversation was successfum
97   // setup data for I2C Transmission
98   ADC_act_data_int = (ADC_act_data_int >> AVERAGING_SHIFT); //
          averaging (devide by AVERAGING SHIFT)
99   ADC_data_LByte = (unsigned char) ADC_act_data_int; // put data
          into bytes for I2C transmission
100   ADC_data_HByte = ((unsigned char) (ADC_act_data_int >> 8)) | (
          Resistance_Range << 2); // plus add the range of the
          previous measurement to the data
101
102   /* *** check and set measurement range (res-only) *** */
103   if (ManRange <= 5){   // manual selection
104   Resistance_Range = ManRange; // set resistance range
105   }
106   else {   // automatic selection
107   // check if previous measurement was done in the optimal range
          , otherwise change the range for the next measurement

```
108        if( (ADC_act_data_int >= 900) & (Resistance_Range <= 4 ) & (
             Resistance_Range >= 0 ) ){
109          Resistance_Range++;
110        }
111        else if( (ADC_act_data_int <= 100) & (Resistance_Range <= 5 )
             & (Resistance_Range >= 1 ) ){
112          Resistance_Range--;
113        }
114      }
115
116      // set Range in Hardware
117      if( Resistance_Range <= 3 ){
118        PORTB &= ~((1 << PORTB5)|(1 << PORTB4)); // deactivate PB.4
             and PB.5
119        PORTA = (1 << (4+Resistance_Range)); // activate corresponding
             PA.x
120      }
121      else if (Resistance_Range == 4){
122        PORTA &= 0x0F;                  // deactivate PA.x
123        PORTB &= ~(1 << PORTB4);        // deactivate PB.4 (1M)
124        PORTB |= (1 << PORTB5);         // activate PB.5 (100k)
125      }
126      else if (Resistance_Range >= 5){
127        Resistance_Range = 5;
128        PORTA &= 0x0F;                  // deactivate PA.x
129        PORTB &= ~(1 << PORTB5);        // deactivate PB.5 (100k)
130        PORTB |= (1 << PORTB4);         // activate PB.4 (1M)
131      }
132      // in lowest measurement range, ATTiny Gain (x8) is needed,
             else (x1)!
133      if(Resistance_Range == 0){
134        ADCSRB |= (1 << GSEL);          // select 8x gain
135      }
136      else{
137        ADCSRB &= ~(1 << GSEL);         // select 1x gain
138      }
139      /* ******************************************************** */
140    }
141  }
142  else{      // invalid communication (not all 24 bits read)
143    /* ignore present shiftregister value -- do nothing */
144  }
145  USISR = 0xF0;            // Clear possible happened TWI flags
146  USICR |= 0x80;          // reactivate TWI Interrupts
147 }
148
149 /* *** reset shiftregister and caliper-bitcounter *** */
150 shift_data = 0;          // reset shift register
151 caliper_count = 0;       // reset caliper-bitcount
152 PORTB1_bit = 1;          // set indicator bit
```

```
153 }
154
155 void External0_ISR() org IVT_ADDR_INT0 {
156
157  /* ***
158  at every falling edge of the Caliper_Clock Signal the present
         Caliper_Data is shifted into the shift_data array. This is done
         for all 24 bits of the Datastream.
159  Furthermore, on first entry the acquisition of the ADC is done.
160  *** */
161
162  // deactivate USI Interrupts
163  USICR &= 0x3F;
164  USISR = 0xF0;
165
166  // on first external interrupt start ADC conversion and reset
         indicator bit
167  if (PINB1_bit) {
168   ADC_act_data_int = 0;        // reset ADC data
169   ADC_conversion_count = 0;   // reset ADC conversion counter
170  }
171
172  PORTB1_bit = 0;               // reset inicator Bit
173  TCNT0L = 0;                   // reset Timer on external interrupt
174
175  /* *** Read caliper levels *** */
176  shift_data = shift_data >> 1; // shift present data by 1 bit
177  caliper_count++;             // increment caliper-bit-count
178
179  // read databit from caliper-data line
180  // insert new data bit (datastream is inverted, therefore !Port3_bit
         )
181  if( !PINB3_bit){
182      shift_data |= 0x80000000;
183  }
184  else { /* leave the 0, which was inserted by the shift operation */
         }
185
186  /* *** Start ADC Acquisition + Averaging measures *** */
187    if(ADC_conversion_count < ADC_AVERAGING){ // on last entry don't
           start a new conversion
188      ADCSRA |= (1 << ADSC);          // Start new ADC Conversion
189    }
190 }
191
192 void ADC0_ISR() org IVT_ADDR_ADC {
193
194  /* ***
195  on finished ADC conversion save the data into a temp variable. All
         acquired values are added and afterwards divided by the number of
```

```
          values (done in Timer0 Interrupt). Therefore an average value
          can be achieved
196    *** */
197
198    ADC_conversion_count++;
199    ADC_LByte_int = (unsigned int) ADCL; // save ADC data
200    ADC_HByte_int = (unsigned int) ADCH;
201    ADC_act_data_int += ((ADC_LByte_int) | (ADC_HByte_int << 8)); // put
          ADC data into temp variable
202    }
```

6.2.3 NXT Communication Routine

This routine is responsible for handling the I^2C communication with the NXT. If the NXT initializes a communication by sending a start-condition, controller-internal electronics trigger the start-condition-interrupt *USI_Start_Condition_ISR*. This routine checks if the data acquisition is idle, reconfigures the TWI-interface and initializes the communication state machine. Immediately after the start-condition the NXT sends one byte containing the ID of the slave he wants to talk to and the data-direction bit. This data is picked up by a controller-internal shift register which triggers the TWI-overflow interrupt *USI_TWI_Overflow_State* after the reception of one byte. If the received ID matches the slaves ID, the slave knows that he is meant. The next step depends on the requested data direction, master wants data or master wants to send data. If the master requests data the slave flushes the corresponding transmit buffers and copies the current position and ADC data in the transmit register. If the master wants to send data to the slave, the slave flushes the corresponding receive buffers. Lastly the slave acknowledges the correct reception of the byte by calling the macro *SET_USI_TO_SEND_ACK()*.

The rest of the communication depends on if the master wants to send or receive data to or from the slave respectively. In principle these two options are rather similar and mostly handled by the controller itself. The important thing to know is, that after eight clocks from the master the overflow-interrupt is called. If the master sends data to the slave the received byte is always located in the receive buffer and has to be handled. In this case only one byte containing the wanted measurement range is received and therefore copied from the TWI-data-register *USIDR* into the variable *ManRange*. If the master requests the last data-point, only the transmit-buffer-index has to be incremented, since the data was already copied into the transmit-(ring)-buffer after the slave was addressed. The indexed byte is then copied into the TWI-data-register and automatically transferred to the NXT.

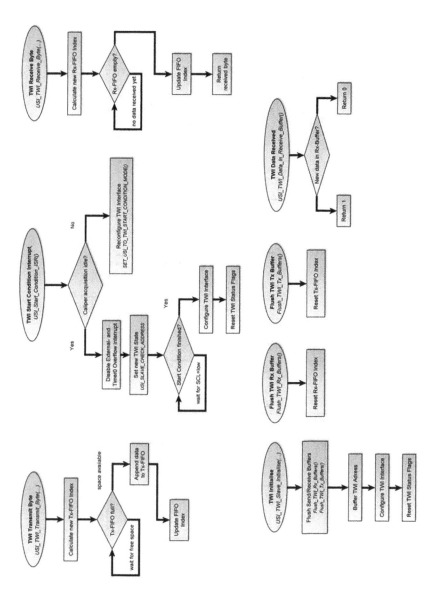

Figure 6.3 – Flowchart of the TWI communication routines to the NXT.

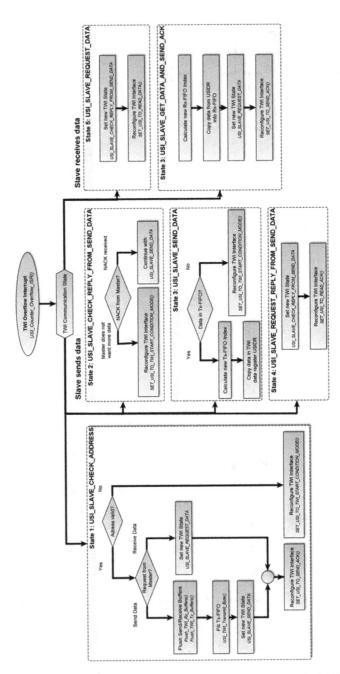

Figure 6.4 – Flowchart of the TWI communication routines to the NXT.

6.2.3.1 USI_TWI_Slave.h

```
 1 /* ********************************************************************
 2 *
 3 * Somap Corporation
 4 *
 5 * File          : USI_TWI_Slave.h
 6 * Compiler      : MicroC PRO
 7 * Revision      : Revision: 5.00
 8 * Date          : Date: Monday, November 12, 2012
 9 * Updated by    : Author: Richard Moser
10 *
11 * Supported devices : ATTiny261
12 *
13 *
14 * Description   : Header file TWI/I2C communication with LEGO NXT2.0
15 *                 The ATTiny acts as Slave in this configuration
16 *                 based on the ATMEL AVR312 Application Note.
17 *
18 ********************************************************************/
19
20 void      USI_TWI_Slave_Initialise( unsigned char );
21 void      USI_TWI_Transmit_Byte( unsigned char );
22 unsigned char USI_TWI_Receive_Byte( void );
23 unsigned char USI_TWI_Data_In_Receive_Buffer( void );
24 void      Timer_Init(void);
25
26 #define TRUE         1
27 #define FALSE        0
28
29 typedef   unsigned char   uint8_t;
30
31 /////////////////////////////////////////////////////////////////////
32 ///////////////// Driver Buffer Definitions ////////////////////////
33 /////////////////////////////////////////////////////////////////////
34 // 1,2,4,8,16,32,64,128 or 256 bytes are allowed buffer sizes
35
36 #define TWI_RX_BUFFER_SIZE (8)
37 #define TWI_RX_BUFFER_MASK ( TWI_RX_BUFFER_SIZE - 1 )
38
39 #if ( TWI_RX_BUFFER_SIZE & TWI_RX_BUFFER_MASK )
40      #error TWI RX buffer size is not a power of 2
41 #endif
42
43 // 1,2,4,8,16,32,64,128 or 256 bytes are allowed buffer sizes
44
45 #define TWI_TX_BUFFER_SIZE (8)
46 #define TWI_TX_BUFFER_MASK ( TWI_TX_BUFFER_SIZE - 1 )
47
48 #if ( TWI_TX_BUFFER_SIZE & TWI_TX_BUFFER_MASK )
49      #error TWI TX buffer size is not a power of 2
```

```
50 #endif
51
52 #define USI_SLAVE_CHECK_ADDRESS          (0x00)
53 #define USI_SLAVE_SEND_DATA              (0x01)
54 #define USI_SLAVE_REQUEST_REPLY_FROM_SEND_DATA (0x02)
55 #define USI_SLAVE_CHECK_REPLY_FROM_SEND_DATA (0x03)
56 #define USI_SLAVE_REQUEST_DATA           (0x04)
57 #define USI_SLAVE_GET_DATA_AND_SEND_ACK  (0x05)
58
59 #define DDR_USI        DDRB
60 #define PORT_USI       PORTB
61 #define PIN_USI        PINB
62 #define PORT_USI_SDA   PORTB0
63 #define PORT_USI_SCL   PORTB2
64 #define PIN_USI_SDA    PINB0
65 #define PIN_USI_SCL    PINB2
66 #define USI_START_COND_INT USISIF
67 #define USI_START_VECTOR IVT_ADDR_USI_START
68 #define USI_OVERFLOW_VECTOR IVT_ADDR_USI_OVF
69
70 // Often used functions implemented as macros
71 #define SET_USI_TO_SEND_ACK()                                    \
72 {                                                                \
73   USIDR  = 0;              /* Prepare ACK */                     \
74   DDR_USI |= (1<<PORT_USI_SDA); /* Set SDA as output */          \
75                 /* Clear all flags, except Start Cond */ \
76   USISR  = (0<<USI_START_COND_INT)|(1<<USIOIF)|(1<<USIPF)|(1<<USIDC)| \
     \
77           (0x0E<<USICNT0); /* set USI counter to shift 1 bit. */ \
78 }
79
80 #define SET_USI_TO_READ_ACK()                                    \
81 {                                                                \
82   DDR_USI &= ~(1<<PORT_USI_SDA); /* Set SDA as intput */         \
83   USIDR  = 0;              /* Prepare ACK   */                   \
84   /* Clear all flags, except Start Cond */                       \
85   USISR  = (0<<USI_START_COND_INT)|(1<<USIOIF)|(1<<USIPF)|(1<<USIDC)| \
     \
86          (0x0E<<USICNT0); /* set USI counter to shift 1 bit. */ \
87 }
88
89 #define SET_USI_TO_TWI_START_CONDITION_MODE()                    \
90 {                                                                \
91   /* Enable Start Condition Interrupt. Disable Overflow Interrupt.*/ \
92   /* Set USI in Two-wire mode. No USI Counter overflow hold. */ \
93   /* Shift Register Clock Source = External, positive edge */ \
94   USICR  = (1<<USISIE)|(0<<USIOIE)|                              \
95           (1<<USIWM1)|(0<<USIWM0)|                               \
96          (1<<USICS1)|(0<<USICS0)|(0<<USICLK)|                   \
97          (0<<USITC);                                            \
```

```
98  /* Clear all flags, except Start Cond */                        \
99  USISR  = (0<<USI_START_COND_INT)|(1<<USIOIF)|(1<<USIPF)|(1<<USIDC)|
        \
100              (0x0<<USICNTO);                                     \
101  INTFO_bit = 1;             /* Clear EXTO Flag */               \
102  INTO_bit = 1;              /* reactivate caliper acquisition */ \
103  TCNTOL = 0;                /* Initialize Timer */              \
104 }
105
106 #define SET_USI_TO_SEND_DATA()                                   \
107 {                                                                \
108  DDR_USI |= (1<<PORT_USI_SDA); /* Set SDA as output */          \
109  /* Clear all flags, except Start Cond */                        \
110  USISR  = (0<<USI_START_COND_INT)|(1<<USIOIF)|(1<<USIPF)|(1<<USIDC)|
        \
111              (0x0<<USICNTO); /* set USI to shift out 8 bits */ \
112 }
113
114 #define SET_USI_TO_READ_DATA()                                   \
115 {                                                                \
116  DDR_USI &= ~(1<<PORT_USI_SDA); /* Set SDA as input */          \
117  /* Clear all flags, except Start Cond */                        \
118  USISR  = (0<<USI_START_COND_INT)|(1<<USIOIF)|(1<<USIPF)|(1<<USIDC)|
        \
119              (0x0<<USICNTO); /* set USI to shift out 8 bits */ \
120 }
```

6.2.3.2 USI_TWI_Slave.c

```
1 /* ********************************************************************
2 *
3 * Somap Corporation
4 *
5 * File         : USI_TWI_Slave.c
6 * Compiler     : MicroC PRO
7 * Revision     : Revision: 5.00
8 * Date         : Date: Monday, November 12, 2012
9 * Updated by   : Author: Richard Moser
10 *
11 * Supported devices : ATTiny261
12 *
13 *
14 * Description   : Source file TWI/I2C communication with LEGO NXT2.0
15 *                 The ATTiny acts as Slave in this configuration
16 *                 based on the ATMEL AVR312 Application Note.
17 *
18 ********************************************************************/
19
20 #include "USI_TWI_Slave.h"
```

```
21 #include "Lego_Caliper_Slave.h"
22
23 /* Static Variables
24  */
25
26 static unsigned char TWI_slaveAddress;
27 static volatile unsigned char USI_TWI_Overflow_State;
28
29 /* Local variables
30  */
31 static uint8_t TWI_RxBuf[TWI_RX_BUFFER_SIZE];
32 static volatile uint8_t TWI_RxHead;
33 static volatile uint8_t TWI_RxTail;
34
35 static uint8_t TWI_TxBuf[TWI_TX_BUFFER_SIZE];
36 static volatile uint8_t TWI_TxHead;
37 static volatile uint8_t TWI_TxTail;
38
39 unsigned int act_caliper_data = 0;
40 char *cpointer;
41
42 /* *** Flushes the TWI Buffers *** */
43
44 void Flush_TWI_Rx_Buffers(void)
45 {
46    TWI_RxTail = 0;
47    TWI_RxHead = 0;
48 }
49 void Flush_TWI_Tx_Buffers(void)
50 {
51    TWI_TxTail = 0;
52    TWI_TxHead = 0;
53 }
54 /* *** USI_TWI functions *** */
55
56 // Initialise USI for TWI Slave mode.
57 void USI_TWI_Slave_Initialise( unsigned char TWI_ownAddress )
58 {
59   Flush_TWI_Tx_Buffers();                    // Reset RX and TX buffers
60   Flush_TWI_Rx_Buffers();                    // Reset RX and TX buffers
61
62   TWI_slaveAddress = TWI_ownAddress;         // Set Slave-Address
63
64   PORT_USI |= (1<<PORT_USI_SCL);             // Set SCL high
65   PORT_USI |= (1<<PORT_USI_SDA);             // Set SDA high
66   DDR_USI |= (1<<PORT_USI_SCL);              // Set SCL as output
67   DDR_USI &= ~(1<<PORT_USI_SDA);             // Set SDA as input
68   USICR = (1<<USISIE)|(0<<USIOIE)|           // Enable Start Condition
69        Interrupt, Disable Overflow Interrupt.
```

```
69              (1<<USIWM1)|(0<<USIWM0)|      // Set USI in Two-wire mode.
                No USI Counter overflow prior to first Start Condition
                (potentail failure)
70              (1<<USICS1)|(0<<USICS0)|(0<<USICLK)| // Shift Register
                Clock Source = External, positive edge
71              (0<<USITC);
72   USISR  = 0xF0;                           // Clear all flags and reset
        overflow counter
73 }
74
75 //Puts data in the transmission buffer, Waits if buffer is full.
76 void USI_TWI_Transmit_Byte( unsigned char dataTX )
77 {
78    unsigned char tmphead;
79
80    tmphead = ( TWI_TxHead + 1 ) & TWI_TX_BUFFER_MASK; // Calculate
         buffer index.
81    while ( tmphead == TWI_TxTail );          // Wait for free space in
         buffer.
82    TWI_TxBuf[tmphead] = dataTX;              // Store data in buffer.
83    TWI_TxHead = tmphead;                     // Store new index.
84 }
85
86 //Returns a byte from the receive buffer. Waits if buffer is empty.
87 unsigned char USI_TWI_Receive_Byte( void )
88 {
89    unsigned char tmptail;
90    unsigned char tmpRxTail;                  // Temporary variable to
         store volatile
91    tmpRxTail = TWI_RxTail;                   // Not necessary, but
         prevents warnings
92    while ( TWI_RxHead == tmpRxTail );
93    tmptail = ( TWI_RxTail + 1 ) & TWI_RX_BUFFER_MASK; // Calculate
         buffer index
94    TWI_RxTail = tmptail;                     // Store new index
95    return TWI_RxBuf[tmptail];                // Return data from the
         buffer.
96 }
97
98 //Check if there is data in the receive buffer.
99 unsigned char USI_TWI_Data_In_Receive_Buffer( void )
100 {
101    unsigned char tmpRxTail;                 // Temporary variable to
          store volatile
102    tmpRxTail = TWI_RxTail;                  // Not necessary, but
          prevents warnings
103    return ( TWI_RxHead != tmpRxTail );      // Return 0 (FALSE) if the
          receive buffer is empty.
104 }
105
```

```
106 /* USI start condition ISR
107  * Detects the USI_TWI Start Condition and intialises the USI
108  * for reception of the "TWI Address" packet.
109  */
110
111 void USI_Start_Condition_ISR(void) iv USI_START_VECTOR {
112
113    unsigned char tmpUSISR;              // Temporary variable to
           store volatile
114
115    // Set default starting conditions for new TWI package
116    if ((PINB1_bit != 0)) {              // check if caliper
           acquisition is busy
117    // if caliper acquisition is idle, deactivate it by disabling the
118    // INTO
119    INTO_bit = 0;
120    tmpUSISR = USISR;                    // Not necessary, but
           prevents warnings
121
122    USI_TWI_Overflow_State = USI_SLAVE_CHECK_ADDRESS; // Set next step
123    DDR_USI &= ~(1<<PORT_USI_SDA);       // Set SDA as input
124
125    while ( (PIN_USI & (1<<PORT_USI_SCL)) ); // Wait for SCL to go low
           to ensure the "Start Condition" has completed.
126
127    USICR =  (1<<USISIE)|(1<<USIOIE)|    // Enable Overflow and
           Start Condition Interrupt. (Keep StartCondInt to detect RESTART
           )
128             (1<<USIWM1)|(1<<USIWM0)|    // Set USI in Two-wire mode

129             (1<<USICS1)|(0<<USICS0)|(0<<USICLK)| // Shift Register
                   Clock Source = External, positive edge
130             (0<<USITC);
131    // Clear flags
132    // Set USI to sample 8 bits i.e. count 16 external pin toggles and
           clear flags.
133    USISR  = 0xF0;
134
135    }
136    else {
137      SET_USI_TO_TWI_START_CONDITION_MODE();
138    }
139 }
140
141 /* USI counter overflow ISR
142  * Handels all the comunication. Is disabled only when waiting
143  * for new Start Condition, or if caliper acquisition is busy.
144  */
145
146 void USI_Counter_Overflow_ISR(void) iv USI_OVERFLOW_VECTOR {
```

```
147
148   unsigned char tmpTxTail; // Temporary variables to store volatiles
149   unsigned char tmpUSIDR;
150
151   switch (USI_TWI_Overflow_State)
152   {
153
154     // *** Address mode ***
155     // Check address and send ACK (and next USI_SLAVE_SEND_DATA) if OK
            , else reset USI.
156     case USI_SLAVE_CHECK_ADDRESS:
157       if ( (( USIDR>>1 ) == TWI_slaveAddress) ) // check if Adress is
              valid
158       {
159         if ( USIDR & 0x01 ) {                      // Master wants to
                receive data
160           USI_TWI_Overflow_State = USI_SLAVE_SEND_DATA; // Slave sends
                data
161
162           Flush_TWI_Tx_Buffers();                 // On fresh
                communication cycle Flush Buffers and copy data to TX
                Buffer
163
164           // Firstly send Caliper-Data
165           act_caliper_data = get_act_caliper_value();
166           cpointer = &act_caliper_data;
167           USI_TWI_Transmit_Byte( (*cpointer) );
168           cpointer++;
169           USI_TWI_Transmit_Byte( (*cpointer) );
170           // Secondly send ADC-Data
171           USI_TWI_Transmit_Byte( get_act_ADC_LByte() );
172           USI_TWI_Transmit_Byte( get_act_ADC_HByte() );
173         }
174         else {
175           USI_TWI_Overflow_State = USI_SLAVE_REQUEST_DATA; // Slave
                receives data
176         }
177         SET_USI_TO_SEND_ACK();                     // Send Acknowledge
                bit
178       }
179       else
180       {
181         SET_USI_TO_TWI_START_CONDITION_MODE();   // not this slave is
                meant
182       }
183       break;
184
185     // *** Master write data mode ***
186
187     // Check reply and goto USI_SLAVE_SEND_DATA if OK, else reset USI.
```

```
188    case USI_SLAVE_CHECK_REPLY_FROM_SEND_DATA:
189     if ( USIDR ) // If NACK, the master does not want more data.
190     {
191       SET_USI_TO_TWI_START_CONDITION_MODE();
192       return;
193     }
194     // From here we just drop straight into USI_SLAVE_SEND_DATA if
          the master sent an ACK

196    // Copy data from buffer to USIDR and set USI to shift byte. Next
          USI_SLAVE_REQUEST_REPLY_FROM_SEND_DATA
197    case USI_SLAVE_SEND_DATA:

199     // Get data from Buffer
200     tmpTxTail = TWI_TxTail;       // Not necessary, but prevents
          warnings
201     if ( TWI_TxHead != tmpTxTail ) // checks if data to send is
          present
202     {
203       TWI_TxTail = ( TWI_TxTail + 1 ) & TWI_TX_BUFFER_MASK;
204       USIDR = TWI_TxBuf[TWI_TxTail];
205     }
206     else                          // If the buffer is empty then:
207     {
208       SET_USI_TO_TWI_START_CONDITION_MODE();
209       return;
210     }
211     USI_TWI_Overflow_State = USI_SLAVE_REQUEST_REPLY_FROM_SEND_DATA;
212     SET_USI_TO_SEND_DATA();
213     break;

215    // Set USI to sample reply from master. Next
          USI_SLAVE_CHECK_REPLY_FROM_SEND_DATA
216    case USI_SLAVE_REQUEST_REPLY_FROM_SEND_DATA:
217     USI_TWI_Overflow_State = USI_SLAVE_CHECK_REPLY_FROM_SEND_DATA;
218     SET_USI_TO_READ_ACK();
219     break;

221    // *** Master read data mode ***
222    // Set USI to sample data from master. Next
          USI_SLAVE_GET_DATA_AND_SEND_ACK.
223    case USI_SLAVE_REQUEST_DATA:
224     USI_TWI_Overflow_State = USI_SLAVE_GET_DATA_AND_SEND_ACK;
225     SET_USI_TO_READ_DATA();
226     break;

228    // Copy data from USIDR and send ACK. Next USI_SLAVE_REQUEST_DATA
229    case USI_SLAVE_GET_DATA_AND_SEND_ACK:
230     // Put data into Buffer
231     tmpUSIDR = USIDR;         // Not necessary, but prevents warnings
```

```
232    TWI_RxHead = ( TWI_RxHead + 1 ) & TWI_RX_BUFFER_MASK;
233    TWI_RxBuf[TWI_RxHead] = tmpUSIDR;
234
235    USI_TWI_Overflow_State = USI_SLAVE_REQUEST_DATA;
236    SET_USI_TO_SEND_ACK();
237    break;
238  }
239 }
```

6.3 Sample Grip CAD

This section contains the CADs for the clamping contraption. Figure
6.5 and 6.6 show the assembled and exploded assembly drawing. On the
following pages the engineering drawings can be found. For assembly in
addition to the aluminium parts two M4 threaded rods with l = 110mm,
one M3 threaded rod with l = 75mm, eight M4 nuts, three M3 nuts and
four 12mm M3 Torx screws are necessary.

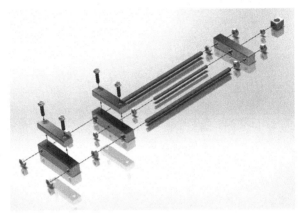

Figure 6.5 – Exploded assembly drawing of the clamping contraption.

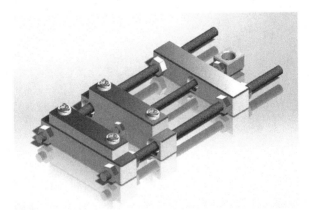

Figure 6.6 – Assembled clamping contraption.

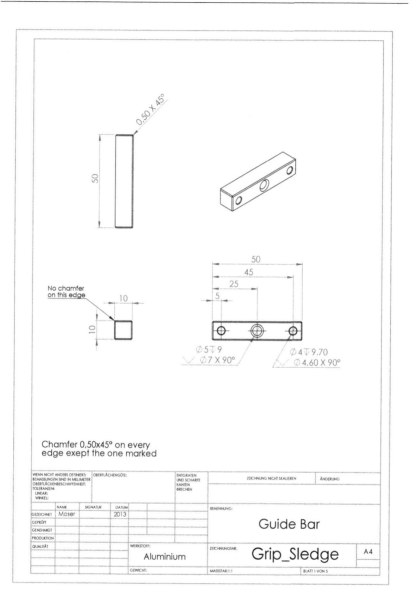

No chamfer
on this edge

0.50 X 45°

50

10

10

50
45
25
5

⌀5⍖9
⌄7 X 90°

⌀4⍖9.70
⌀4.60 X 90°

Chamfer 0,50x45° on every
edge exept the one marked

WENN NICHT ANDERS DEFINIERT: BEMASSUNGEN SIND IN MILLIMETER OBERFLÄCHENBESCHAFFENHEIT: TOLERANZEN: LINEAR: WINKEL:	OBERFLÄCHENGÜTE:			ENTGRATEN UND SCHARFE KANTEN BRECHEN	ZEICHNUNG NICHT SKALIEREN	ÄNDERUNG	
	NAME	SIGNATUR	DATUM		BENENNUNG:		
GEZEICHNET	Moser		2013				
GEPRÜFT					Guide Bar		
GENEHMIGT							
PRODUKTION							
QUALITÄT			WERKSTOFF:		ZEICHNUNGSNR.		
			Aluminium		Grip_Sledge		A4
			GEWICHT:		MASSSTAB:1:1		BLATT 1 VON 5

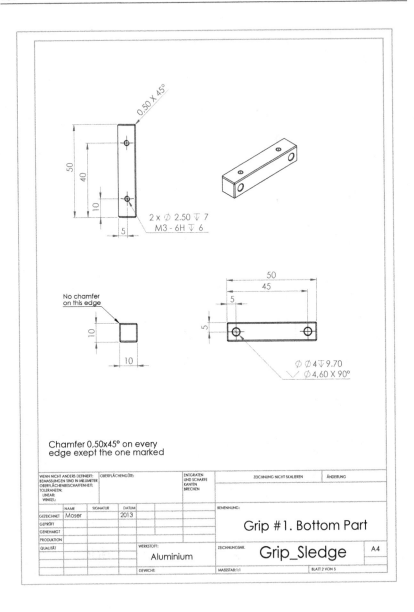

2 x ⌀ 2.50 ⍌ 7
M3 - 6H ⍌ 6

No chamfer
on this edge

⌀ ⌀4 ⍌ 9.70
⌵ ⌀4,60 X 90°

Chamfer 0,50x45° on every
edge exept the one marked

WENN NICHT ANDERS DEFINIERT: BEMASSUNGEN SIND IN MILLIMETER OBERFLÄCHENBESCHAFFENHEIT: TOLERANZEN: LINEAR: WINKEL:	OBERFLÄCHENGÜTE:		ENTGRATEN UND SCHARFE KANTEN BRECHEN	ZEICHNUNG NICHT SKALIEREN	ÄNDERUNG	
	NAME	SIGNATUR	DATUM	BENENNUNG:		
GEZEICHNET	Moser		2013			
GEPRÜFT				**Grip #1. Bottom Part**		
GENEHMIGT						
PRODUKTION						
QUALITÄT			WERKSTOFF:	ZEICHNUNGSNR.		
			Aluminium	**Grip_Sledge**		A4
			GEWICHT:	MASSSTAB:1:1	BLATT 2 VON 5	

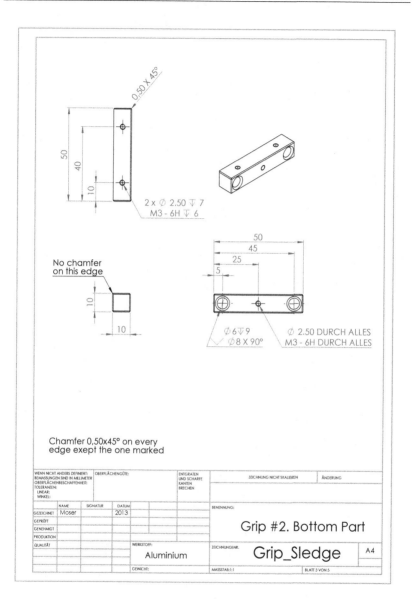

0,50 X 45°

50
40
10

2 x Ø 2.50 ⊽ 7
M3 - 6H ⊽ 6

No chamfer
on this edge

10

10

50
45
25
5

Ø 6 ⊽ 9
⊽ Ø 8 X 90°

Ø 2.50 DURCH ALLES
M3 - 6H DURCH ALLES

Chamfer 0,50x45° on every
edge exept the one marked

WENN NICHT ANDERS DEFINIERT: BEMASSUNGEN SIND IN MILLIMETER OBERFLÄCHENBESCHAFFENHEIT: TOLERANZEN: LINEAR: WINKEL:	OBERFLÄCHENGÜTE:		ENTGRATEN UND SCHARFE KANTEN BRECHEN		ZEICHNUNG NICHT SKALIEREN	ÄNDERUNG
	NAME	SIGNATUR	DATUM		BENENNUNG:	
GEZEICHNET	Moser		2013			
GEPRÜFT					**Grip #2. Bottom Part**	
GENEHMIGT						
PRODUKTION						
QUALITÄT		WERKSTOFF:		ZEICHNUNGSNR.	**Grip_Sledge**	A4
		Aluminium				
		GEWICHT:		MASSSTAB:1:1		BLATT 3 VON 5

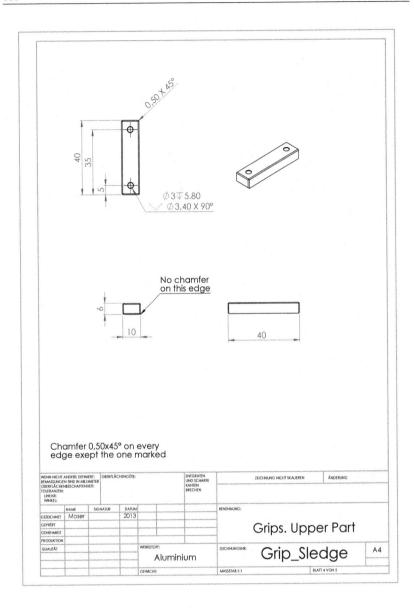

No chamfer
on this edge

Chamfer 0,50x45° on every
edge exept the one marked

WENN NICHT ANDERS DEFINIERT:	OBERFLÄCHENGÜTE:		ENTGRATEN UND SCHARFE KANTEN BRECHEN	ZEICHNUNG NICHT SKALIEREN	ÄNDERUNG
BEMASSUNGEN SIND IN MILLIMETER OBERFLÄCHENBESCHAFFENHEIT: TOLERANZEN: LINEAR: WINKEL:					

	NAME	SIGNATUR	DATUM		BENENNUNG:	
GEZEICHNET	Moser		2013			
GEPRÜFT					**Grips. Upper Part**	
GENEHMIGT						
PRODUKTION						
QUALITÄT			WERKSTOFF:		ZEICHNUNGSNR.	
			Aluminium		**Grip_Sledge**	A4
			GEWICHT:		MASSSTAB:1:1	BLATT 4 VON 5

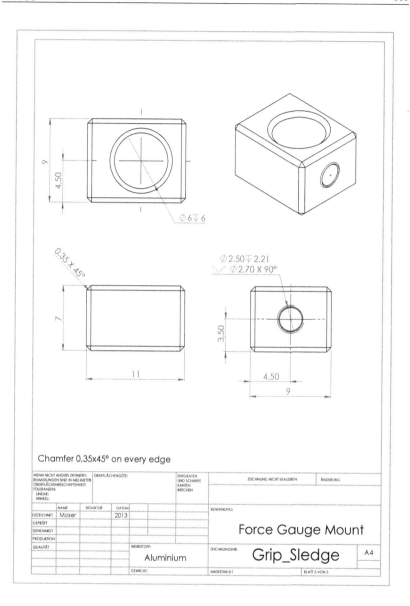

Chamfer 0,35x45° on every edge

WENN NICHT ANDERS DEFINIERT: BEMASSUNGEN SIND IN MILLIMETER OBERFLÄCHENBESCHAFFENHEIT: TOLERANZEN: LINEAR: WINKEL:	OBERFLÄCHENGÜTE:			ENTGRATEN UND SCHARFE KANTEN BRECHEN	ZEICHNUNG NICHT SKALIEREN	ÄNDERUNG
	NAME	SIGNATUR	DATUM		BENENNUNG:	
GEZEICHNET	Moser		2013			
GEPRÜFT					**Force Gauge Mount**	
GENEHMIGT						
PRODUKTION						
QUALITÄT			WERKSTOFF:		ZEICHNUNGSNR.	
			Aluminium		**Grip_Sledge**	A4
			GEWICHT:		MASSSTAB:5:1	BLATT 5 VON 5

References

[1] J. A. Rogers, T. Someya, and Y. Huang, "Materials and mechanics for stretchable electronics," *Science*, vol. 327, no. 5973, pp. 1603–1607, 2010.

[2] Plastic Logic Ltd. (2013, Apr.) Flexible electrophoretic plastic displays. [Online]. Available: http://www.plasticlogic.com/

[3] M. Kaltenbrunner, M. S. White, E. D. Głowacki, T. Sekitani, T. Someya, N. S. Sariciftci, and S. Bauer, "Ultrathin and lightweight organic solar cells with high flexibility," *Nature Communications*, vol. 3, no. 770, 2012.

[4] R. Kaltseis, C. Keplinger, R. Baumgartner, M. Kaltenbrunner, T. Li, P. Mächler, R. Schwödiauer, Z. Suo, and S. Bauer, "Method for measuring energy generation and efficiency of dielectric elastomer generators," *Applied Physics Letters*, vol. 99, no. 162904, 2011.

[5] IMEC International. (2013, Apr.) Stretchable circuit fabricated using a proprietary stretchable molded interconnect. [Online]. Available: http://www.imec.be/

[6] I. M. Graz and S. P. Lacour, "Flexible pentacene organic thin film transistor circuits fabricated directly onto elastic silicone membranes," *Applied Physics Letters*, vol. 95, no. 243305, 2009.

[7] G. Kettlgruber, M. Kaltenbrunner, C. M. Siket, R. Moser, I. M. Graz, R. Schwödiauer, and S. Bauer, "Intrinsically stretchable and rechargeable batteries for self-powered stretchable electronics," *J. Mater. Chem. A*, vol. 1, no. 18, pp. 5505–5508, 2013.

[8] I. M. Graz, "Anschmiegsame Elektronik. Mechanik der Makroelektronik," *Physik in unserer Zeit*, vol. 40, no. 5, pp. 243–249, 2009.

[9] D.-H. Kim, N. Lu, R. Ma, Y.-S. Kim, R.-H. Kim, S. Wang, J. Wu, S. M. Won, H. Tao, A. Islam *et al.*, "Epidermal electronics," *Science*, vol. 333, no. 6044, pp. 838–843, 2011.

[10] TestResources Inc. (2013, Apr.) Universal Testing Machines, Zwick 314 Series. [Online]. Available: http://www.testresources.net/

[11] W. Demtrdöer, *Experimentalphysik 1: Mechanik und Wärme.* Springer Verlag, 2005.

[12] G. R. Strobl, *The physics of polymers: concepts for understanding their structures and behavior.* Springer Verlag, 1997.

[13] J. H. Weiner, *Statistical mechanics of elasticity.* DoverPublications, 2002.

[14] L. Treloar, *The physics of rubber elasticity.* Oxford University Press, 2005.

[15] L. D. Landau and E. M. Lifshitz, *Lehrbuch der theoretischen Physik Band 7: Elastizitätstheorie.* Akademie Verlag, 1991.

[16] P. Kelly. (2013, Mar.) Solid mechanics lecture notes, Department of Engineering Science, University of Auckland. [Online]. Available: http://homepages.engineering.auckland.ac.nz/~pkel015/ SolidMechanicsBooks/

[17] R. Ogden, *Non-linear elastic deformations.* Dover Publications, Inc, 1997.

[18] I. M. Graz, *Lecture notes: stretchable electronics, Johannes Kepler University Linz,* 2013.

[19] University of Cambridge, Department of Engineering. (2013, Mar.) Material selection charts. [Online]. Available: http://www-materials. eng.cam.ac.uk/mpsite/physics/introduction/

[20] W. Demtröder, *Experimentalphysik 3: Atome, Moleküle, Festkörper.* Springer Verlag, 2005.

[21] (2013, Apr.) Polyisoprene chain. [Online]. Available: http://www.tennoji-h.oku.ed.jp/tennoji/oka/OCDB/SyntheticRubber/syspolyisoprene-b.gif

[22] R. Bartoletti, LearnNC. (2013, Apr.) Polyisoprene molecule. [Online]. Available: http://www.learnnc.org/lp/pages/4496

[23] Wikipedia Commons. (2013, Apr.) Isoprene molecule. [Online]. Available: http://commons.wikimedia.org/wiki/File:Isoprene-Structure.png

[24] ——. (2013, Apr.) Isoprene molecule. [Online]. Available: http://commons.wikimedia.org/wiki/File:Isoprene-3D-balls-B.png

[25] J. Mark and B. Erman, *Rubberlike elasticity: a molecular primer.* Cambridge University Press, 2007.

[26] M. Mooney, "A theory of large elastic deformation," *Journal of Applied Physics*, vol. 11, no. 9, pp. 582–592, 1940.

[27] R. S. Rivlin and D. Saunders, "Large elastic deformations of isotropic materials. vii. experiments on the deformation of rubber," *Philosophical Transactions of the Royal Society of London. Series A, Mathematical and Physical Sciences*, vol. 243, no. 865, pp. 251–288, 1951.

[28] A. Gent and F. R. Eirich, "Rubber elasticity: basic concepts and behavior," *Science and Technology of Rubber. Academic Press*, pp. 1–21, 1978.

[29] C. Horgan and G. Saccomandi, "A molecular-statistical basis for the Gent constitutive model of rubber elasticity," *Journal of Elasticity*, vol. 68, no. 1, pp. 167–176, 2002.

[30] M. C. Boyce, "Direct comparison of the Gent and the Arruda-Boyce constitutive models of rubber elasticity," *Rubber Chemistry and Technology*, vol. 69, no. 5, pp. 781–785, 1996.

[31] R. Ogden, "Large deformation isotropic elasticity-on the correlation of theory and experiment for incompressible rubberlike solids," *Proceedings of the Royal Society of London. A. Mathematical and Physical Sciences*, vol. 326, no. 1567, pp. 565–584, 1972.

[32] M. C. Boyce and E. M. Arruda, "Constitutive models of rubber elasticity: a review," *Rubber Chemistry and Technology*, vol. 73, no. 3, pp. 504–523, 2000.

[33] LEGO Group. (2013, Mar.) Lego Mindstorms NXT, user guide. [Online]. Available: http://education.lego.com/en-us/preschool-and-school/upper-primary/8plus-mindstorms-education

[34] ——. (2013, Mar.) Official LEGO Education website - The LEGO history. [Online]. Available: http://aboutus.lego.com/en-us/lego-group/the_lego_history/

[35] C. Beland, W. Chan, D. Clarke, R. Park, and M. Trupiano. (2013, Apr.) Lego mindstorms: The structure of an engineering (r)evolution. [Online]. Available: http://web.mit.edu/6.933/www/Fall2000/LegoMindstorms.pdf

[36] LEGO Group. (2013, Mar.) Official LEGO Mindstorms website. [Online]. Available: http://mindstorms.lego.com/en-us/default.aspx

[37] ——. (2013, Mar.) Official LEGO Education website. [Online]. Available: http://education.lego.com/default.aspx?domainredir=www.legoeducation.com

[38] E. Wang, J. LaCombe, and C. Rogers, "Using LEGO bricks to conduct engineering experiments," in *Proceedings of the ASEE Annual conference and exhibition*, 2004.

[39] M. Tully. (2013, Mar.) How The Infinity Project, the LEGO Mindstorms NXT robot and National Instruments LabVIEW software are being combined to transform engineering education in

ireland at fourth grade primary school and upwards. [Online]. Available: http://www.ineer.org/Events/ICEEiCEER2009/full_papers/full_paper_109.pdf

[40] A. Behrens, L. Atorf, R. Schwann, B. Neumann, R. Schnitzler, J. Balle, T. Herold, A. Telle, T. G. Noll, K. Hameyer *et al.*, "MATLAB meets LEGO Mindstorms: A freshman introduction course into practical engineering," *Education, IEEE Transactions on*, vol. 53, no. 2, pp. 306–317, 2010.

[41] D. Cliburn, "An introduction to the Lego Mindstorms," in *Proceedings of the 2006 ASCUE Conference, Myrtle Beach, South Carolina, June 11-15, 2006,*.

[42] J. Ringwood, K. Monaghan, and J. Maloco, "Teaching engineering design through LEGO Mindstorms," *European Journal of Engineering Education*, vol. 30, no. 1, pp. 91–104, 2005.

[43] M. B. Vallim, J.-M. Farines, and J. E. Cury, "Practicing engineering in a freshman introductory course," *IEEE Transactions on Education*, vol. 49, no. 1, pp. 74–79, 2006.

[44] R. Saint-Nom and D. Jacoby, "Building the first steps into SP research [signal processing education]," in *Proceedings of the ICASSP international conference on acoustics, speech, and signal processing, 2005.*

[45] D. G. Strange and M. L. Oyen, "Biomimetic bone-like composites fabricated through an automated alternate soaking process," *Acta Biomaterialia*, vol. 7, no. 10, pp. 3586–3594, 2011.

[46] D. Pile, "Optical components: LEGO lightens photonics," *Nature Photonics*, vol. 3, no. 7, pp. 377–378, 2009.

[47] F. Quercioli, B. Tiribilli, A. Mannoni, and S. Acciai, "Optomechanics with LEGO," *Applied Optics*, vol. 37, no. 16, pp. 3408–3416, 1998.

[48] D. R. Albert, M. A. Todt, and H. F. Davis, "A low-cost quantitative absorption spectrophotometer," *Journal of Chemical Education*, vol. 89, no. 11, pp. 1432–1435, 2012.

[49] W. S. Hua, J. R. Hooks, W. J. Wu, and W. C. Wang, "Development of a polymer based fiberoptic magnetostrictive metal detector system," in *International Symposium on Optomechatronic Technologies (ISOT)*. IEEE, 2010, pp. 1–5.

[50] J. A. Olson, C. E. Calderon, P. W. Doolan, E. A. Mengelt, A. B. Ellis, G. C. Lisensky, and D. J. Campbell, "Chemistry with refrigerator magnets: from modeling of nanoscale characterization to composite fabrication," *Journal of Chemical Education*, vol. 76, no. 9, p. 1205, 1999.

[51] K. Knagge and D. Raftery, "Construction and evaluation of a LEGO spectrophotometer for student use," *The Chemical Educator*, vol. 7, no. 6, pp. 371–375, 2002.

[52] A. Heiskanen, V. Coman, N. Kostesha, D. Sabourin, N. Haslett, K. Baronian, L. Gorton, M. Dufva, and J. Emnéus, "Bioelectrochemical probing of intracellular redox processes in living yeast cells – application of redox polymer wiring in a microfluidic environment," *Analytical and Bioanalytical Chemistry*, pp. 1–12, 2013.

[53] T. C. McAlpine, C. Atwood-Stone, T. Brown, and J. F. Lindner, "Tracking the stars, Sun, and Moon to connect with the universe," *American Journal of Physics*, vol. 78, p. 1128, 2010.

[54] LEGO Group. (2013, Mar.) Official LEGO Space website. [Online]. Available: http://legospace.com/en-us/Default.aspx

[55] Peekelectronics. (2013, Mar.) Peekelectronics official webpage. [Online]. Available: http://www.peekelectronics.co.uk

[56] Guinness World Records. (2013, Mar.) Guinness World Records official webpage. [Online]. Available: http://www.guinnessworldrecords.de/

[57] LUGNET. (2013, Mar.) International LEGO Users Group Network webpage. [Online]. Available: http://www.lugnet.com/

[58] M. Gasperi, R. Hempel, L. Villa, and D. Baum, *Extreme Mindstorms: an advanced guide to LEGO Mindstorms*. Apress, 2000.

[59] J. B. Knudsen and M. Loukides, *The unofficial guide to LEGO Mindstorms robots*. O'Reilly, 1999.

[60] P. Wallich, "Mindstorms: not just a kid's toy," *Spectrum, IEEE*, vol. 38, no. 9, pp. 52–57, 2001.

[61] Vernier Software & Technology. (2013, Mar.) Vernier official website. [Online]. Available: www.vernier.com

[62] HiTechnic Products. (2013, Mar.) HiTechnic official website. [Online]. Available: http://www.hitechnic.com/

[63] Dexter Industries. (2013, Mar.) Dexter Industries official website. [Online]. Available: http://www.dexterindustries.com/

[64] Mindsensors.com. (2013, Mar.) Mindsenors.com official website. [Online]. Available: http://www.mindsensors.com/

[65] P. E. Hurbain and M. Gasperi, *Extreme NXT: Extending the LEGO Mindstorms NXT to the next level*. Apress, 2007.

[66] National Instruments Corporation. (2013, Mar.) NI LabView für LEGO Mindstorms. [Online]. Available: http://www.ni.com/academic/mindstorms/d/

[67] M. Noga. (2013, Mar.) BrickOS official webpage. [Online]. Available: http://brickos.sourceforge.net/

[68] D. Baum. (2013, Mar.) NQC - Not Quite C. [Online]. Available: http://bricxcc.sourceforge.net/nqc/

[69] RWTH Aachen University. (2013, Mar.) RWTH - LEGO Mindstorms NXT Toolbox. [Online]. Available: http://www.mindstorms.rwth-aachen.de/

[70] The MathWorks, Inc. (2013, Mar.) LEGO Mindstorms NXT support from MATLAB and Simulink. [Online]. Available: http://www.mathworks.de/academia/lego-mindstorms-nxt-software/

[71] Robomatter, Inc. (2013, Mar.) RobotC, a C programming language for robotics. [Online]. Available: http://www.robotc.net/

[72] LeJos. (2013, Mar.) LeJos, JAVA for LEGO Mindstorms website. [Online]. Available: http://lejos.sourceforge.net/

[73] M. Kossman. (2013, Mar.) LEGO Technic design blog. [Online]. Available: http://technic.lego.com/en-us/designers/blog/default.aspx?date=2/16/2011

[74] Atmel Corporation. (2013, Feb.) Data Sheet ATtiny261, Rev. 2588E 08/10. [Online]. Available: http://www.atmel.com/Images/doc2588.pdf

[75] NXP. (2013, Mar.) UM10204, I2C-bus specification and user manual, Rev. 5: 9 October 2012. [Online]. Available: http://www.nxp.com/documents/user_manual/UM10204.pdf

[76] LEGO Group. (2013, Mar.) Lego Mindstorms NXT, hardware development kit. [Online]. Available: http://mindstorms.lego.com/en-us/support/files/default.aspx#Advanced

[77] Wayne and Layne, LLC. (2013, Mar.) LEGO Mindstorms sensor cable. [Online]. Available: http://www.wayneandlayne.com

[78] Mitutoyo America Corporation. (2013, Mar.) Mitutoyo America official webpage. [Online]. Available: http://www.mitutoyo.com/

[79] I. M. Graz, D. P. Cotton, and S. P. Lacour, "Extended cyclic uniaxial loading of stretchable gold thin-films on elastomeric substrates," *Applied Physics Letters*, vol. 94, no. 7, p. 071902, 2009.

[80] A. P. Robinson, I. Minev, I. M. Graz, and S. P. Lacour, "Microstructured silicone substrate for printable and stretchable metallic films," *Langmuir*, vol. 27, no. 8, p. 4279, 2011.

[81] S. Chung, J. Lee, H. Song, S. Kim, J. Jeong, and Y. Hong, "Inkjet-printed stretchable silver electrode on wave structured elastomeric substrate," *Applied Physics Letters*, vol. 98, no. 15, pp. 153 110–153 110, 2011.

[82] S. Rosset, M. Niklaus, P. Dubois, and H. R. Shea, "Metal ion implantation for the fabrication of stretchable electrodes on elastomers," *Advanced Functional Materials*, vol. 19, no. 3, pp. 470–478, 2009.

[83] Texas Instruments Incorporated. (2013, Mar.) TL431, precision programmable reference. [Online]. Available: http://www.ti.com/lit/ds/symlink/tl431.pdf

[84] ——. (2013, Mar.) LM358, low power dual operational amplifiers. [Online]. Available: http://www.ti.com/lit/ds/symlink/lm158-n.pdf

[85] Fairchild Semiconductor. (2013, Mar.) BC547, NPN epitaxial silicon transistor. [Online]. Available: http://www.fairchildsemi.com/ds/BC/BC547.pdf

[86] ——. (2013, Mar.) FDS9926A, dual n-channel 2.5V specified power trench Mosfet. [Online]. Available: http://www.datasheetcatalog.org/datasheet/fairchild/FDS9926A.pdf

[87] Analog Devices, Inc. (2013, Mar.) Ad623, single-supply, rail-to-rail, low cost instrumentation amplifier. [Online]. Available: http://www.analog.com/en/specialty-amplifiers/instrumentation-amplifiers/ad623/products/product.html

[88] "ISO527-3: Plastics - determination of tensile properties - Part 3: Test conditions for films and sheets." 1996.

[89] J. T. Bauman, *Fatigue, stress, and strain of rubber components: A guide for design engineers.* Carl Hanser GmbH, 2008.

[90] F. Schneider, T. Fellner, J. Wilde, and U. Wallrabe, "Mechanical properties of silicones for MEMS," *Journal of Micromechanics and Microengineering*, vol. 18, no. 6, p. 065008, 2008.

[91] GOM mbH. (2013, Apr.) ARAMIS - Optical 3D deformation analysis. [Online]. Available: http://www.gom.com/metrology-systems/system-overview/aramis.html

[92] ASTM International, "ASTM D412 - 06ae2, standard test methods for vulcanized rubber and thermoplastic elastomers: Tension." 2008.

[93] K. M. Choi and J. A. Rogers, "A photocurable poly (dimethylsiloxane) chemistry designed for soft lithographic molding and printing in the nanometer regime," *Journal of the American Chemical Society*, vol. 125, no. 14, pp. 4060–4061, 2003.

[94] D. Cotton, A. Popel, I. Graz, and S. Lacour, "Photopatterning the mechanical properties of polydimethylsiloxane films," *Journal of Applied Physics*, vol. 109, no. 5, p. 054905, 2011.

[95] R. Huang and L. Anand, "Non-linear mechanical behavior of the elastomer polydimethylsiloxane (pdms) used in the manufacture of microfluidic devices," *Department of Mechanical Engineering, Massachusetts Institute of Technology*, 2005.

[96] Y.-S. Yu and Y.-P. Zhao, "Deformation of PDMS membrane and microcantilever by a water droplet: Comparison between Mooney–Rivlin and linear elastic constitutive models," *Journal of Colloid and Interface Science*, vol. 332, no. 2, pp. 467–476, 2009.

[97] K. Khanafer, A. Duprey, M. Schlicht, and R. Berguer, "Effects of strain rate, mixing ratio, and stress–strain definition on the mechanical behavior of the polydimethylsiloxane (PDMS) material as related to its biological applications," *Biomedical Microdevices*, vol. 11, no. 2, pp. 503–508, 2009.

Printed in the United States
By Bookmasters